Harvey Washington Wiley

Record of Experiments

in the manufacture of sugar from sorghum at Rio Grande, New Jersey; Kenner,

Louisiana; Conway Springs, Douglass, and Sterling, Kansas. 1888

Harvey Washington Wiley

Record of Experiments
in the manufacture of sugar from sorghum at Rio Grande, New Jersey; Kenner, Louisiana; Conway Springs, Douglass, and Sterling, Kansas. 1888

ISBN/EAN: 9783337219284

Printed in Europe, USA, Canada, Australia, Japan

Cover: Foto ©berggeist007 / pixelio.de

More available books at **www.hansebooks.com**

U. S. DEPARTMENT OF AGRICULTURE.

DIVISION OF CHEMISTRY.

BULLETIN   No. 20.

RECORD OF EXPERIMENTS

CONDUCTED BY THE

COMMISSIONER OF AGRICULTURE

IN THE

MANUFACTURE OF SUGAR FROM SORGHUM

AT

RIO GRANDE, NEW JERSEY; KENNER, LOUISIANA; CONWAY SPRINGS, DOUGLASS, AND STERLING, KANSAS.

1888.

BY

H. W. WILEY, Chemist.

————•✦•————

WASHINGTON:
GOVERNMENT PRINTING OFFICE.
1889.

14056—Bull. 20——1

LETTER OF SUBMITTAL.

DECEMBER 19, 1888.

SIR: I have the honor to submit herewith the manuscript of Bulletin No. 20, being the report of experiments in the manufacture of sugar from sorghum conducted by your direction during the season of 1888.

Respectfully,

H. W. WILEY,
Chemist.

Hon. NORMAN J. COLMAN,
Commissioner of Agriculture.

3

EXPERIMENTS IN THE MANUFACTURE OF SUGAR FROM SORGHUM.

ASSIGNMENT OF WORK.

The bill making an appropriation for experiments in the manufacture of sugar did not become a law until the 19th of July, 1888. At that time it was manifestly impossible for the Department to make any arrangements of its own for the conduct of experiments during the present manufacturing season. It was necessary, if any experiments were to be made at all, that they should be arranged for in connection with work already in progress either by individuals, private corporations, or State experiment stations. The following arrangements were therefore made for the experimental work:

(1) A continuation of the experimental work at Rio Grande, N. J., under the direction of Mr. H. A. Hughes.

(2) A series of experiments at Kenner, La., under the direction of Prof. W. C. Stubbs.

(3) Experimental work at Douglass, Kans., under the direction of the Douglass Sugar Company.

(4) Experimental work at Conway Springs, Kans., under the direction of Mr. E. W. Deming.

(5) Experiments in the improvement in the varieties of cane at Sterling, Kans., under the direction of Mr. A. A. Denton.

In addition to the above work arrangements were made for analytical researches under my direction at Douglass, Conway Springs and Sterling, Kans. It was deemed unadvisable at the late date mentioned for the Department to suggest any experimental work or assume any control thereof. Having been authorized to arrange for such work in a manner which seemed most advantageous the following directions were given. The work at Rio Grande was placed exclusively in charge of Mr. H. A. Hughes, to be conducted in such a manner as he saw fit for the benefit of the industry. The work which Mr. Hughes proposed to do was on a small scale, with the ultimate idea of making it possible for farmers and others to manufacture sugar without the expense of apparatus usually considered necessary for that purpose. The results of Mr. Hughes's work have been reported by him further on in this bulletin, and a discussion of them will be given in connection with his report.

Prof. W. C. Stubbs having commenced preparations for experimental work with sorghum at the experiment station at Kenner, he was authorized to complete this work under the auspices of the Department. No instructions in regard to the method of performing the work were sent Professor Stubbs, except to do that which seemed best for the promotion of the industry. His report of the results of the work and the discussion thereof will follow.

The experimental work at Douglass, Kans., was placed under the control of the Douglass Sugar Company. The object was to test thoroughly the method of open diffusion practiced on a small scale by Mr. Hughes, at Rio Grande, and they conducted the work under the general instructions to give that system of diffusion and the apparatus a thorough and impartial test. The general results of the experimental work at the station are given in the report of Mr. Edson, with a discussion of the data there recorded.

The experimental work at Conway Springs consisted in the trial of a new system of preparing the exhausted chips for fuel; and certain new arrangements of apparatus connected with the diffusion battery and of a new system of handling and storing the cane. No specific instructions were given to Mr. Deming in regard to the conduct of the work, but he was left free to use his own judgment in every particular in regard to what was best to be done. Mr. Deming's report and the discussion thereof will follow.

The experimental work at Sterling was of an entirely different order. The Sterling Sugar Company had commenced a thorough examination of all obtainable varieties of the sorghum plant. By an arrangement made with this company, the Department assumed this work in the condition in which it was found the latter part of July and carried it to completion under the supervision of Messrs. Denton and Crampton, whose report and observations thereon will follow.

The following assignment of the chemical force of the division was made for the purpose of securing analytical data of the season's work. Mr. Hughes having expressed an opinion that he could get along independently of any chemical assistance from the Department, no assignment was made to Rio Grande. Mr. Edson was placed in charge of the chemical work at Douglass, assisted by Mr. John L. Fuelling. Prof. E. A. von Schweinitz was placed in charge of the chemical work at Conway Springs, assisted by Mr. Oma Carr. Dr. C. A. Crampton was placed in charge of the work at Sterling, assisted by Mr. Karl P. McElroy.

In the latter part of July I visited the three localities last named, and arranged with the proper persons for the establishment of the laboratories and perfected the arrangements for the chemical control which was desired. In September and October I visited each of the laboratories above mentioned, and spent some days with the chemists in charge in consultation concerning the progress of the work and any changes or alterations therein which seemed necessary. The results of the chemical work in each case will be found in connection with the reports of the respective stations.

EXPERIMENTS AT RIO GRANDE, N. J.

The result of the work at Rio Grande is disappointing in its nature. For some reason the cane grown in that locality has failed to improve, although it appears that it has had the benefit of careful attention and fertilization. There has been upon the whole, as indicated in Bulletin 18, a deterioration of the cane at Rio Grande, the crops which were raised six or seven years ago showing a higher percentage of sucrose than those of the present time. This deterioration has been caused either by admixture of a non-saccharine variety with the seed, by the method of culture, or by the influence of the soil and climate of that locality. I am inclined to attribute much of the depreciation to a fault of the seed; whether or not it has been mixed with broom-corn I am unable to say. The almost total failure of the amber cane at Rio Grande would seem to indicate that some such accident had happened to it. While amber cane in other localities has continued to show a high percentage of sucrose in the juice, at Rio Grande it has become a worthless variety for sugar-making or even the production of sirup. The importance of seed selection is emphasized by this fact, since there is every reason to believe that if seed of the early amber, such as was planted at Rio Grande seven or eight years ago, were again planted in that locality it would produce an equally rich crop of cane. It would be a useless task, however, for any one to attempt the successful manufacture of sugar by any process from juices no richer than those reported by Mr. Hughes during the present year; such canes at best could only make molasses, and that probably of an inferior character. These agricultural results are the more discouraging because of the systematic attempts which have been made at Rio Grande in conjunction with the New Jersey experiment station for the production of a high-grade cane; these are not, however, sufficiently discouraging to justify abandonment of similar attempts in other localities. In respect of the climate at Rio Grande, I can see nothing which would lead me to believe that it is unfavorable to the growth of sorghum. On the other hand, the climatic conditions appear extremely favorable, unless it be true that sorghum will not develop a maximum content of sugar in localities favored with abundant summer rains. Aside from this, the favorable conditions for growth and the practical immunity from early frosts render the locality a most favorable one for the production and manufacture of a crop of sorghum cane. The soil of this locality, it is true, is not naturally as fertile as the soils of Kansas, but with the judicious fertilization which has been practiced, the tonnage per acre has been fully as great, if not greater, at Rio Grande than in most other localities.

In regard to the methods of manufacture employed at this station, it is necessary to speak with some degree of caution. In the report of Mr. Hughes we have, from his stand-point, a brief but graphic description of the method employed. I have never been of the opinion that sugar-making from sorghum could be successfully practiced on a small scale, and the experiments carried on by the Department of Agriculture for two successive seasons at Rio Grande have only served to confirm me in this belief. The nature of the processes employed, the character of machinery required, and the kind of skilled labor needed, all combine to render the manufacture of sugar on a small scale commercially unsuccessful. I do not see any favorable result in this direction from the two years' trial at Rio Grande. For the present manufacturing season Mr. Hughes does not give the total amount of sugar made, except from a portion of the crop, and this is no evidence whatever that its cost has been sufficiently low to enable it to be put upon the market in competition with other sugars. I should have been glad had the result been otherwise, for the successful inauguration of an era of sugar-making conducted by farmers would have been a great blessing to vast agricultural regions.

In regard to the machinery employed my opinion has already been expressed. I have said repeatedly, both in official publications and in other places, that I regarded the system of cutting and preparing the cane devised by Mr. Hughes, and now in use in every sorghum factory in the United States and in at least one cane-sugar factory, as the very best which has yet been invented. I have long been convinced that for the extraction of sugar from cane of both kinds the greater the degree of comminution of the chips the more successful the process will be. The system of double shredding inaugurated by Mr. Hughes during the past season tends to secure this end. It was in this direction also that I urged last year for sugar-cane the construction of a shredding-machine on the principle of the shredder built by the Newell Universal Mill Company of New York, for the purpose of preparing the pieces of cane properly for the diffusion battery. This shredder I suggested should be furnished with very fine steel knives, of the general pattern of the shredder now in use, with short cylinders of large diameter, driven at a very much higher rate of speed. Last year I suggested to Mr. Fiske, the inventor of the machine above mentioned, the advisability of building such a machine in duplicate for the purpose of reducing the cane to as fine pieces as possible. The advantage of such a shredder as this over the one used by Mr. Hughes would be principally in its greater strength, and in the assurance that it could be run for days, and perhaps a whole season through, without any necessity for repairs. It is of the highest importance that the apparatus for cutting and pulping the cane should be as effective as possible and built in two sets, so that if one should be out of order the second could still be used.

In regard to the system of diffusion practiced at the Rio Grande station, and described in Bulletin 18, further experience only leads me to emphasize what was said in that bulletin, viz:

The defects of the system were both mechanical and chemical.

The mechanical difficulty is the same as that which attends all methods of diffusion in which the cane chips are moved instead of the diffusion liquors. From a mechanical point of view it is far easier and more economical to move a liquid in a series of vessels than a mass of chips. In the Hughes system the whole mass of chips under going diffusion, together with adhering liquor, and baskets and suspending apparatus are lifted vertically a distance of several feet, varying with the depth of the diffusion, tanks, every few minutes. The mechanical energy required to do this work is enormous, and with large batteries the process would prove almost impossible.

The truth of this view will be further illustrated in the report of the Douglass Sugar Company. For very small batteries working only a few tons a day this system might possibly be employed, but I doubt even then if it could be economically worked. This opinion of mine, as will be seen, is at total variance with that expressed by Mr. Hughes, and those who propose to become practically interested in the matter will have to decide upon the merits of the two systems of diffusion after a personal investigation.

Mr. Hubert Edson, who has had two years' experience with the open system of diffusion, made the following statements relating thereto in the Lousiana Planter and Sugar Manufacturer of December 1, 1888. His report refers to the battery used at Douglass, Kans., during the season of 1888:

The battery was built from plans secured directly from Mr. Hughes, and with one or two slight changes was worked throughout the season. The main battery consisted of ten cells, open at the top to admit the baskets in which the chips were placed for diffusion. These baskets, made of strong boiler-iron, were attached to the arms of a crane, which was raised, rotated, and lowered till the requisite number of immersions was obtained. Besides these ten cells there was an extra one of the same dimensions placed just outside and within reach of the arms from the large crane. This arrangement was intended to secure a dense diffusion juice, allowing, as the diffusion progressed, the heaviest juice from two of the cells of the main battery to be drawn into the outside cell, and which there received two baskets of fresh chips before being emptied.

This manner of operating the battery will, it is claimed by the inventor, give a juice almost as dense as a corresponding mill juice. In my opinion, however, no greater advantage is secured by the eleventh cell being outside the main battery than by the same number arranged in regular order. Certainly, at Douglass, the results claimed by the inventor were not even approximated. The outside cell also entailed an extra amount of labor in transferring the basket from the small crane, to which it was attached during its immersion, to the large crane of the main battery.

So much for the manner of working the battery. Now for the things that are of actual value to the sugar planters, the results obtained, and the expenses incident to such results.

Machinery of any kind to be effective should require a minimum of human labor. Let us see how the Hughes battery compares with the ordinary form. At Douglass the battery was designed to work 100 tons of cane daily, and to do this at least eight men were necessary to shift the baskets to their different places. Half of this number would run a close battery and find the work easier, since they would have no baskets weighing 1,000 pounds each to handle.

Besides this manual labor the whole ten baskets had to be raised every time one was filled or emptied. A large hydraulic pump is used for this work and of itself requires more power than is necessary to run a battery of closed cells. This extra power and labor would not necessarily condemn the apparatus if such superior results were obtained as to overcome the expense. But instead of this, exactly the reverse was accomplished. Not much better extraction was secured than is obtained by the ordinary cane-mill of Louisiana, and this only with a dilution of nearly 50 per cent., causing an extra expense of no small amount for evaporation. Then, also, the quality of the juice obtained was extremely poor. The almost constant exposure to the air and especially in iron vessels blackened it to such a degree that no good sugars could be made from it. Clarification was nearly impossible with any of the ordinary re-agents in the sugar-house. This was extremely unfortunate in Kansas, as the greatest profits are made on material sold to the home market.

Full reports of the chemical work at Rio Grande are contained in Bulletin 51, New Jersey Experiment Station.

EXPERIMENTS AT KENNER, LA.

As has been mentioned before, Prof. W. C. Stubbs was placed in charge of the experiments which were arranged for in connection with the Louisiana sugar-experiment station at Kenner and the stations at Baton Rouge and Calhoun. For two previous seasons Professor Stubbs had made extensive experiments with sorghum, which are fully reported in the bulletins of the Louisiana experiment station and in Bulletin No. 18 of this division. A study of the analytical data of the three years' work in Louisiana shows in an emphatic way the peculiarities of sorghum which have rendered so difficult the successful inauguration of sugar-making from that plant. The great variations in the content of sucrose in the juices of the plant, its susceptibility to injury by storms and other unforeseen causes, are strikingly set forth in the analytical figures which follow. In my opinion the production of a variety of sorghum-cane suitable to the soil and climate of the sugar lands of Louisiana will be a work of no small difficulty. From the results of the work already done, and especially during the last year, an account of which is contained in the appended report of Professor Stubbs, it is clearly seen that a season which has produced a sugar-cane very rich in sucrose in the State of Louisiana has produced a sorghum crop which is absolutely worthless for sugar-making for commercial purposes. Another point illustrated by the report is brought out in the reference to the past work of the station, in which, although a cane was produced whose juice was reasonably rich in sucrose, its practical working in the sugar factory was found most difficult. In the report this is ascribed to the presence of large quantities of dextrine or dextrine-like bodies supposed to be derived from the starch originally present in the juice. It is the opinion of Professor Stubbs that starch and sucrose are developed in the sorghum *pari passu*. In this case it would be found that the direct polarization of a sorghum juice rich in sugar would show apparently a much higher content of sucrose than was actually present, since dextrine and its allied bodies are much more strongly dextro gyratory than sucrose. The points developed by the experiments may be summarized as follows :

(1) Sorghum cane develops sometimes in Louisiana a juice containing a very high percentage of sucrose, but combined with other bodies which render its separation from the juice difficult.

(2) The occurrence of a wet summer attended by the severe wind-storms which are so common in that locality prevent the development of a high sucrose content in the growing sorghum.

(3) The possible utilization of sugar machinery for a longer manu-facturing season is one of the chief inducements in the sugar-cane regions for the cultivation of sorghum as a sugar-producing plant.

(4) Delay in working the cane after cutting is not as dangerous as has been supposed.

It will be understood that these are conclusions which I have drawn from reading Professor Stubbs's report, and are not formulated in the above manner by himself.

The results of attempts to grow sorghum for sugar-making purposes on the low sugar-lands of Louisiana, in my opinion, are not highly en-couraging to the belief that these lands and their climate are the best suited in the United States for the production of sorghum, as Professor Stubbs says. On the other hand, I believe there are few localities in the United States, where sorghum grows at all, in which a better crop for sugar-making purposes can not be produced. Experience has shown that the dry climate of southern and western Kansas produces the most uniform crop of sorghum for sugar-making purposes, while the data of Professor Stubbs, which follow, show that the Louisiana product, for the present year at least, is about the poorest on record. One point, how-ever, should be borne in mind, viz, that the course of experiment pur-sued by the Louisiana experiment station is the one which is best suited for the rapid development of every possibility of sorghum culture in that State. The experimental trials which are made with sorghum will show both its weak and strong points, and in the wide variation which the plant shows there will doubtless be some variety produced or found which will be best suited to the peculiar conditions which obtain in that locality. The soil and climatic conditions of the northern part of the State, where cotton is now grown, will probably be found better suited to the production of sorghum than those of the present sugar-producing localities. I feel quite sure that the expectation expressed by Professor Stubbs of being able to realize under certain conditions as much as 120 to 125 pounds of sugar from sorghum cane may be fully met under favorable circumstances; but it would still remain to be demonstrated that this yield could be reasonably expected from year to year, or even occasionally, on a large scale. The subsequent experiments which are promised by Professor Stubbs at the Louisiana station will doubtless set at rest, in a few years, all these questions, and demonstrate to the sugar makers of Louisiana just what can be expected from sorghum as an adjunct to their great industry.

EXPERIMENTS AT CONWAY SPRINGS.

In the reports of Messrs. Deming and von Schweinitz which follow, together with the analytical tables, much interesting information may be found in regard to the sorghum-sugar industry in Kansas. The successful continuation of the work at Fort Scott has encourged the belief in the possibility of a speedy establishment of a sorghum-sugar industry in Kansas on a large scale. The unfortunate financial outcome of the work at Conway Springs shows that much is yet to be learned by those entering upon this industry before success can be confidently predicted. A discussion of the chemical data collected at Conway Springs will be found in connection with the analytical tables. It is proper to say here, however, that the sorghum juices of the crop grown at Conway Springs show a higher content of sucrose than any large crop which has ever before been produced in the United States. This high content of sucrose which appeared in the crop after the middle of September, as indicated by the analysis of the juices, was continued until the close of the working season in November. The samples of chips taken from the cells of the battery showed in their juices a high content of sucrose uniformly; much higher, in fact, than would be indicated by the output of sugar. One reason, doubtless, for this was the exceptionally dry season diminishing the content of water in the cane and thus increasing the percentage of sucrose in the juice. This fact, though not established by the determination of the fiber in the cane, is plainly indicated by two other facts developed by the analytical work, viz, the diminished extraction when using the small mill at the same pressure as the season progressed and the high content of total solids in the juices. The output of sugar was evidently diminished by the character of the water used in diffusion, but that would be unable to account for the small yield of crystallizable sugar obtained with juices of the richness of those worked. Experiments made by boiling a solution of pure sugar with the water used in diffusion at Conway Springs proved that the presence of a large amount of gypsum did not tend to increase the inversion of sucrose; that it may, however, have interfered with the crystallization of the sucrose is a fact which can scarcely be denied. The actual output of sugar at Conway Springs, in my opinion, would have been considerably larger had pure water been employed in the diffusion battery; nevertheless, the important fact remains that the yield of crystallizable sugar was wholly disproportional to the richness of the juices worked, showing that the high ratio of sucrose was not obtained at the expense of the solids not sugar in the juices. In other words, it appears that a cane whose juice is normal in quantity, say about 90 per cent. of the total weight, and having a content of sugar equal to 10 per cent., with total solids at 16 per cent., will yield fully as much, if not more, sugar than a cane whose juice is abnormal, say not more than 80 per cent. of the total weight, with 12 per cent. of sucrose and 18 or 20 per cent. of total solids. Another important fact developed by a study of the data obtained at Conway Springs is the persistence of the sugar content in the juice after the cane was fully ripened. In localities where considerable moisture may be expected in the soil as a result of frequent rains during the manufacturing season it has been noticed that there is a rapid deterioration of the juices, beginning a short time after complete maturation. This has been especially noticed in the experience at the Rio Grande station. It has also been noticed by all careful observers of sorghum grown in ordinary localities. The inspissation of the juices by the natural causes of an extremely dry climate appears to protect the sugar from this destruction. This is a point of the greatest interest to sorghum-growers, to whom the preservation of the sugar in the juice for a reasonable length of time is a matter of the greatest consideration. In the process of diffusion this thickening of the juice entails no loss, although if milling were used for expressing the juice the loss would be a most serious one. The above explanation of the character of the juice at Conway Springs is offered with some degree of hesitation, since I am fully aware of the danger of drawing conclusions in sorghum work from a too limited number of observations.

The manufacturing operations at Conway Springs were greatly hindered by faults in the machinery, which could scarcely be avoided when the short time allowed for the manufacture and erection of the same is considered. Instead of taking three months for the erection of a sugar factory, a whole year is none too long a time, and disaster, for at least one year, is certain to attend attempts to erect such machinery in the time allowed at Conway Springs.

What is needed now in the sorghum-sugar industry is the manufacture of sugar at a rate which will enable the manufacturer to compete with sugar from other parts of the world. A great step in this direction will be secured when the kind of machinery which has been pointed out by the investigations of the Department as necessary to success shall be constructed by skilled machinists and erected by skilled engineers, with time enough at their disposal to finish their work before the manufacturing season begins. Some further remarks on this subject will be made in another place.

From a commercial point of view, the results of the work at Conway Springs are wholly disappointing. To the person, however, who will take pains to inform himself in regard to the conditions which there obtained, many points of encouragement will be found in spite of the financial failure of the first season's work.

EXPERIMENTS AT DOUGLASS, KANS.

The practical experiments carried on at Douglass consisted in a thorough trial of the open system of diffusion (the Hughes system) to test its fitness for working on a large scale. For the details of the construction of the battery I refer to the report of Mr. Edson. In regard to its working in general, I may say that it was a total failure, both as to economy of power and success of extraction. The financial difficulties which were met with by the company during the year were attributed largely to the use of this battery. The evaporating apparatus in use at Douglass was of first-class quality and arranged in a practical manner. The system of clarification tanks, double effects, and strike pan was as good as could be desired for sugar-making purposes. Had the company adopted the system of diffusion erected by the Department at Fort Scott, there is every reason to believe that even during the first season it would have paid all expenses and made a reasonable profit. The attempt to introduce a new and untried system on a large scale shows the danger which too often besets the introduction of a new enterprise. The promoters of such an enterprise, not satisfied with what has been accomplished, attempt to follow new paths, which often lead to unknown and unwished-for localities. It is best in any enterprise to accept what has been proved of value and not jeopardize the success of a commercial undertaking by introducing in its place a kind of experiment, which, failing, would destroy all prospects of success. As will be seen by the analytical tables accompanying the Douglass report, the crop was of fair quality, showing about the average percentage of sucrose developed in Kansas during the last two or three years. The soil on which most of the crop was raised was somewhat richer in vegetable matter and contained less sand than the soil at Conway Springs. The climatic conditions of the two places were so nearly identical as to make apparently but little difference; yet it must be conceded that at Douglass the hot dry winds produced less effect than at Conway Springs. There did not appear to be the same drying up of the juice, which may account to some extent for the percentage of sucrose therein being less. The agricultural results, however, were of the most encouraging nature, showing that in this locality a crop of sorghum cane can be grown which, with proper treatment, may be expected to yield from 80 to 90 pounds of sugar per ton of clean cane. Not only were the actual results rendered unfavorable by the kind of battery employed, but, aside from this, for some reason the centrifugals used proved to be wholly inadequate to the severe task imposed upon them. The drying of sorghum sugar is at best a difficult task, and only the best approved centrifugal apparatus should ever be employed for this purpose. Had the battery at Douglass worked successfully much delay would have been experienced in the manufacture of the crop by the imperfections above noted in the centrifugal machines.

EXPERIMENTS AT STERLING, KANS.

At the very beginning of my connection with the experiments in the manufacture of sugar from sorghum I realized the importance of improving the quality of the cane to be used. In Bulletin No. 3, page 107, I made the following statements:

The future success of the industry depends on the following conditions, viz:

(1) A careful selection and improvement of the seed with a view of increasing the proportion of sucrose.

(2) A definition of geographical limits of successful culture and manufacture.

(3) A better method of purifying the juices.

(4) A more complete separation of the sugar from the canes.

(5) A more complete separation of the sugar from the molasses.

(6) A systematic utilization of the by products.

(7) A careful nutrition and improvement of the soil.

IMPROVEMENT BY SEED SELECTION.

I am fully convinced that the Government should undertake the experiments which have in view the increase of the ratio of sucrose to the other substances in the juice. These experiments, to be valuable, must continue under proper scientific direction for a number of years. The cost will be so great that a private citizen will hardly be willing to undertake the expense.

The history of the improvement in the sugar beet should be sufficient to encourage all similar efforts with sorghum.

The original forage beet, from which the sugar beet has been developed, contained only 5 or 6 per cent. of sucrose. The sugar beet now will average 10 per cent.* of sucrose. It seems to me that a few years of careful selection may secure a similar improvement in sorghum.

It would be a long step toward the solution of the problem to secure a sorghum that would average, field with field, 12 per cent. sucrose and only 2 per cent. of other sugars, and with such cane the great difficulty would be to make sirup and not sugar. Those varieties and individuals of each variety of cane which show the best analytical results should be carefully selected for seed, and this selection continued until accidental variations become hereditary qualities in harmony with the well-known principles of descent.

If these experiments in selection could be made in different parts of the country, and especially the various agricultural stations and colleges, they would have additional value and force. In a country whose soil and climate are as diversified as in this, results obtained in one locality are not always reliable for another.

If some unity of action could in this way be established among those engaged in agricultural research, much time and labor would be saved and more valuable results be obtained.

In a summary of the methods which I had advocated for the improvement of the sorghum plant, I said in an address before the National Sugar Growers' Association in Saint Louis, in February, 1887:

Finally, our experiments have taught us that after all the mechanical difficulties which have been enumerated in the manufacture of sugar from sorghum have been overcome, the industry can not become commercially successful until the scientific agronomist succeeds in producing a sorghum plant with a reasonably high and uniform content of sucrose and a minimum of other substances. This work is peculiarly the function of our agricultural experiment stations. In beet-sugar-producing countries the production of the seed for planting is a distinct branch of the industry. So,

---

* In the six years that have passed since the above was written the sugar beet has been still further improved and its mean percentage of sucrose now amounts to perhaps 12.

too, it must be with sorghum. A careful scientific selection of the seeds of those plants showing the best sugar-producing qualities, their proper planting and cultivation, a wise choice of locality and soil, a proper appreciation of the best methods of culture, these are all factors which must be taken into consideration in the successful solution of the problem.

It was with this purpose in view that I made the arrangements with the Sterling Sirup Company by which the Department assumed control of the experiments which they had commenced in the cultivation of different varieties of sorghum. At the time this arrangement was made, viz, in the latter part of July, Mr. A. A. Denton was already in charge thereof for the Sterling Sirup Company, and he was appointed to continue in general charge under the direction of the Department. It was arranged with Mr. Denton that the general line of research should be such as is indicated in the above statements of the purposes in view. The chemists who were sent to take charge of the analytical work were instructed to co-operate with Mr. Denton in such a way as to secure favorable results and to make such suggestions as might seem valuable in the details of the work. Mr. Denton was requested to make a general study of the growth of the different varieties and of the habits of each one with reference to its fitness as a sugar plant. The most promising individuals of each variety were to be selected for experimental purposes, and those showing the highest content of sucrose with the lowest content of other substances were to be preserved for future planting. The able manner in which Mr. Denton accomplished this work, assisted by the chemists of the Department, will be found in his detailed report. I regard it of the highest importance to the future success of the industry that the line of work thus commenced by the Department should be continued.

One great difficulty with which we have to contend is in the character of the appropriations made for the experimental work. I have called attention to this difficulty in former reports, and I wish to emphasize the matter here. The fiscal year in all Government affairs begins on the 1st of July. For investigations in agriculture no more unfortunate beginning of the year could be selected. On the 1st of July it is too late to commence experiments for that season; if these experiments be postponed till the next season arrangements can be made for their continuation only up to the 1st of the next July, and thus they have to be stopped before they are well begun. The difficulty is extremely manifest in the present instance. The wisdom and value of continuing the experiments at Sterling last year will be denied by no one. Abundant funds are left over from the present year's appropriation to continue the experiments for another season ; it is, however, unwise to make any arrangements for such work, since no part of it, except that which will be let out by contract, could be continued after the 1st of July, 1889. You thus find your hands tied, as it were, by the unfortunate disposition of the experimental year which has to begin and end with the fiscal year. To avoid this difficulty, which has been one of the greatest causes of the disasters which have attended our experiments with sorghum, I earnestly recommend that all appropriations for field and manufacturing experiments in agricultural matters be made to take effect from the 1st of January each year instead of the 1st of July.

POINTS TO BE CONSIDERED IN BUILDING A FACTORY.

It is of the utmost importance, both for the individuals and the industry, that intending investors in the sugar business should carefully consider the problem presented to them in all its forms. Failure is not only a personal calamity but a public one in that it deters capital from investment in an industry which, properly pursued, gives promise of a fair interest on the money invested.

Soil and climate.—The importance of soil and climate has already been discussed. In the light of present experience it must be conceded that a soil and climate similar to those of southern and western Kansas are best suited to the culture of sorghum for sugar-making purposes. Further investigations may show that Texas and Louisiana present equally as favorable conditions, but this yet awaits demonstration. Conditions approximately similar to those mentioned can doubtless be found in Arkansas, Tennessee, North Carolina and other localities. The expectations which were entertained and positively advocated a few years ago of the establishment of a successful sorghum industry in the great maize fields of the country must now be definitely abandoned. He who would now advise the building of a sorghum-sugar factory in northern Illinois, Indiana, Iowa, or Wisconsin would either betray his ignorance or his malignity. A season of manufacture, reasonably certain for sixty days, is an essential condition to success in the manufacture of sorghum sugar. Early frosts falling on cane still immature, or a freezing temperature on ripe cane followed by warm weather, are alike fatal to a favorable issue of the attempt to make sugar. Sober and careful men will not be misled by the claims of the enthusiast, by the making of a few thousand pounds of sugar in Minnesota, by the graining of whole barrels of molasses in Iowa. Four or five million acres of land will produce all the sugar this country can consume for many years and these acres should be located where the climatic conditions are most favorable. During the past season sorghum cane matured as far north as Topeka, but in 1886 the cane crop at Fort Scott was ruined by a heavy frost on the 29th of September, and in 1885 a like misfortune happened at Ottawa, Kans., on the 4th of October. These interesting facts show that these points are on the extreme northern limits of safety for sorghum-sugar making, and the region of success will be found to the south and west of them.

Natural fertility of soil must also be considered as well as favorable climate. The sandy pine lands of North Carolina can not hope to compete with the rich prairies of southwestern Kansas and the Indian Territory. Indeed, in my opinion, the last-named locality should it ever be opened to white settlers, is destined to be the great center of the sorghum-sugar industry; nevertheless, those who plant the virgin soils of this great southwestern empire must remember that to always take and never give will tire the most patient soils, and a just return should be annually made to the willing fields. A judicious fertilization, rotation of crops, and rest will not only preserve the natural fertility of the fields but give even a richer return in the improved quality of the cane and the greater tonnage secured. Perhaps the most sensible solution of the problem of the disposition of the waste chips will be found in re-. turning them to the soil. These chips have a positive manurial value in the nitrogen they contain, while their merely physical effect on the soil may prove of the highest importance.

Water supply.—The misfortunes which have attended many attempts in the manufacture of sugar by diffusion by reason of an imperfect or insufficient water supply are a sufficient warning on this subject to the careful student. Not only should the water supply be abundant and easily accessible, but the portion of it at least which is to be used in the battery should be as pure as possible. The presence of carbonate of lime and some other carbonates in water is not injurious, but the evil effects of a large amount of other kinds of mineral matter are shown in the data from Conway Springs. When the supply of water is insufficient it has been customary to use ponds for receiving the waste from the factory, so that it may be used again. This method is applicable if care be taken to prevent organic matters, scums, etc., from entering the water supply. In case this precaution is not taken the operator of the factory may find himself in the condition in which the Department was placed in its first experiments at Ottawa and Fort Scott in being compelled to use water foul and putrescent. It is scarcely safe to rely upon a well for a supply of water, especially if it has to be sunk to any depth. Where pumping machinery must be placed many feet below the surface, as in the cramped condition which attends its erection in a well, serious difficulties may arise from the machinery getting out of order, and a great loss of energy may ensue from the necessity of lifting the water to a great height. In all cases where it is possible a running stream of water should be selected for the supply, and the factory should be placed conveniently near its banks. The importance of this matter is emphasized the more when it is considered that the most favorable localities for sugar making, as indicated by the present state of our knowledge, are situated in regions where the water supply is notably deficient. Yet it must be admitted that even in southern and western Kansas it will not be difficult to find localities for the erection of sugar factories where the water supply is certain and abundant. In the light of past experience it is not probable that any further mistakes will be made in this direction. Careful estimates should be made of the quantity of water required, and absolute certainty should be secured of the supply of that amount of water, and even of a much greater amount in cases of emergency. The only safety will be found in some such plan as this.

Proximity of cane fields.—Another point which must be taken into consideration in the location of a factory is the distance which the cane is to be transported. This is a matter which of course the farmers raising the cane are more interested in than the proprietors of the factory, when the cane is grown by contract. With good roads, in a level country, it is easy to draw from 1½ to 2 tons of field cane at each load. The average price which is paid for such cane at the present time is $2 per ton. It is evident that at a given distance, varying according to the price of teams and labor in each locality, the cost of transportation would equal the total receipts for the cane; in this case the farmer would have nothing left to pay for the raising of the cane and profit. Evidently true economy, from an agricultural point of view, would require the cane to be grown as near the factory as possible. It would be well, indeed, if all the cane could be grown within a radius of 1 mile from the factory. This would give, in round numbers, 2,000 acres tributary to a factory. With an ordinary season this ought to produce 20,000 tons of cane. The lengthening of the radius of this circle by one-half mile would give the greatest distance to be hauled 1½ miles, thus vastly increasing the surface tributary to the central factory. It is true that at the present time farmers are easily found who are willing to draw their cane 4, 5, and even 6 miles, but this condition of affairs can not be continued when the business is fully established and the factories in sharp competition with each other. In case the exhausted chips are to be returned to the soil as fertilizer the importance of a centrally located factory, as described, is doubly emphasized.

Fuel.—A cheap and abundant supply of fuel is not less important than the raw material to be manufactured into sugar. As far as the sorghum-sugar industry is concerned the coal which is used for fuel is transported almost exclusively by rail. In locating a factory, therefore, both for convenience of shipping the product and for receiving a supply of fuel, it should be placed sufficiently near a railway line to enable it to be connected therewith by a switch. It is better, however, that the switch should be of some considerable length than that the water supply should be remote or the cane in distant fields.

The problem of burning the exhausted chips has not yet been successfully solved, and I doubt very much whether it will be.* Save the softening which the chips undergo in the process of diffusion the difficulty of expressing the water from them is as great as that of expressing the juice from fresh chips. Thus to dry the chips sufficiently to make them economical for fuel would require a vast expenditure of power, which would hardly be supplied by the increased supply of steam generated by their combustion. Experiments during the seasons 1887–'88 at Magnolia Plantation, Louisiana, showed that an ordinary cane-mill was poorly adapted to the pressure of exhausted cane chips. The feeding of the

---

* Since this was written further experiments are more favorable to the possibility of economically using the chips for fuel.

mill was difficult, and the amount of fuel produced seemed wholly dis-proportional to the expense of preparing it. It has been proposed to try the process used for extracting the water from beet pulp for the purpose of drying sorghum chips. There is nothing whatever in the experience of the beet sugar factories to warrant the belief that such a process would render the chips sufficiently dry to burn. Although I would not be considered as discouraging any further attempts in the direction of preparing sorghum chips for fuel, I must be allowed to ex-press the belief that for some time to come coal must be chiefly re-lied upon.

If the chips are to be successfully burned in the future we may make up our mind, that it will have to be done by previous pressure in mills which in all their appointments shall be as strong and efficient as those which have been in use for expressing the juice from cane. It can not be hoped that these chips will be made sufficiently dry by exposing them to the sun, and in artificial desiccation the amount of fuel required would be almost as great as that used in the evaporation of the original juice. It is claimed that at Wonopringo, in Java, as reported in the New Orleans Item of December 16, 1888, the Fives-Lille Company has succeeded in drying the chips by passing them through two powerful three roll mills, and that the chips thus dried do not contain more than 55 per cent. of moisture and burn readily in an automatic furnace in-vented by Godillot. If it be assumed that 100 pounds of chips contain 10 pounds of combustible matter it is seen that nearly 80 pounds of water will have to be expressed therefrom before they are fit for fuel. I am doubtful whether such a process will prove profitable save in countries where fuel is very dear, as it is in Java and Cuba.

Cost of factory.—It is on almost universal experience that the actual cost of a sugar factory is underestimated by those who undertake its erection. Many of the disasters which have attended the manufacture of sorghum sugar have been due to miscalculation of the cost of the apparatus necessary for the purpose. It is the part of wisdom to avoid mistakes of this kind, and before undertaking the erection of a factory to fully understand the amount of outlay which will be required. The cost of a factory will, of course, vary according to its capacity and the character of the machinery and building erected. In my opinion there is little economy in using cheap machinery, hastily and poorly put to-gether. Success is more likely to be obtained by using the very best machinery which has been devised for sugar-making purposes, and erecting it in a lasting and substantial manner. The economy which is secured in operating such machinery far exceeds that which would be obtained by erecting a cheaper plant. The character of the building must also be taken into consideration; it should be sufficiently large to allow a proper disposition of all parts of the machinery without crowd-ing, and sufficiently strong to afford a proper support for such portions thereof as may rest upon it. Due regard should also be paid to risks of fire, and that portion of the factory especially exposed to such dangers should be made as nearly as possible fire-proof. The plans and specifications for all the machinery should be carefully prepared under the direction of a competent engineer and architect, and the machinery furnished by manufacturing firms whose experience and reputation are a guaranty of the excellence of their work. For a complete factory, capable of working 200 tons per day, the cost may be estimated at $60,000 for a minimum and $100,000 for a maximum, the difference being caused by the elaborateness of the work. This may seem a large sum, but it is highly important that intending investors should know the magnitude of the undertaking which they propose. An estimate which exceeds the actual outlay by $10,000 will be far more satisfactory to all parties concerned than one which falls short of it by the same amount.

Technical and chemical control.—The manufacture of sugar from sorghum is no mysterious process known only to one or two persons, as attempts have been made to establish; nevertheless it must be understood that without experience in the manufacture of sugar the most competent engineer may fail. It is best, therefore, that intending investors understand this beforehand that they may be able to secure some one to take charge of the manufacture of sugar who thoroughly understands the needs of the business and has had some experience in the conduct thereof. Perhaps there are not more than fifteen or twenty such men now in the United States, but their number will be largely increased within a short time. It would seem, therefore, that the number of factories which could be successfully operated in the next year or two is limited, and this fact should be taken into careful consideration by those intending to invest money in the business. An intelligent young man of good education, with quick perceptions and of industrious habits, would be able in one year, working in a sorghum-sugar factory, to obtain a knowledge which would enable him to take charge of a factory, with some degree of success, on his own responsibility. One object which the Department has had in view in its experiments has been in having them open, not only to public inspection, but to careful technical study, to such persons as chose to make the attempt. It is to be regretted that at least one company, who through the courtesy of the Commissioner of Agriculture was permitted to use a large amount of machinery belonging to the Department, has so far forgotten its obligations to the public as to refuse permission for a technical study and report on its operations during the past year. Public property is devoted to a poor purpose when used in such a manner.

The importance of chemical control of the manufacturing work is so evident that I need not dwell upon it long. The vagaries of the sorghum plant are so pronounced as to require the careful supervision of the chemist at all times. In localities not far removed differences in the character of the sorghum are most marked, as illustrated by the data obtained at Conway Springs and Douglass, Kans., during the past year. To determine the fitness of the cane for the manufacture of sugar, control the workings of the factory, and find and remove the sources of loss in the sugar-house, are duties which can be committed only to the chemist. For many years, at least, this chemical supervision will be necessary, and its utility will always continue.

PROGRESS OF DIFFUSION WITH SUGAR-CANE.

Two plantations are using the process of diffusion during the present season for the extraction of sugar from sugar-cane. These are Sugar Land plantation of Colonel Cunningham, in Texas, and the Magnolia plantation of Governor Warmoth, in Louisiana. The latest reports from the Sugar Land plantation I find in the Item of December 15, 1888. At that time it is reported that over 2,000,000 pounds of sugar had been made and that the diffusion battery was working up from 300 to 350 tons of cane a day. It is also reported that an average of 194 pounds of sugar is made per ton. From the analyses of the cane reported in the Item of November 28, 1888, it appears that the juice has about 12 per cent. of crystallizable sugar. The success of the operations seems to be fully assured.

The working of the battery at Magnolia is also satisfactory. The analysis of the cane shows that it is extremely rich in sugar. In the Item of December 4 it is reported that the juice contained 13.7 to 16.6 per cent. of sugar. A polarization had been made showing as high as 19.2 per cent.

Under date of December 9, Mr. G. L. Spencer writes as follows:

Diffusion is working to everyone's satisfaction. We have had a great many delays, almost all of which were caused by the Yaryan quadruple-effect pan. Governor Warmoth had the apparatus overhauled this morning and found that the exhaust-pipe from the pump opens into the second effect, making a pressure-pan of this when working with more than 3 or 4 pounds of steam. This defect has been remedied and we hope everything will be all right now. The cutter gave a great deal of trouble at first, so much that we thought it would be necessary to abandon it. Finally two holes cut in the side of the casing opposite the cutting disk relieved it, so now it is working well. We can cut a cell of chips averaging 2,864 pounds in seven and a half minutes. The dilution will probably surprise you. I intended starting with a dilution of 33 per cent., but by a mistake in measurement I started with 50 per cent. With 50 per cent. dilution we left from .28 to .70 sucrose in the chip juice. I gradually reduced the dilution until it dropped to 14.8 per cent., leaving about .70 to 1 per cent. of sucrose in the exhausted chip juices. We have finally commenced running with a dilution of 21 per cent., leaving .42 per cent. of sucrose in the exhausted chip juices. With pulped cane, such as Hughes's apparatus gives, I would be willing to guaranty a dilution of only 18 per cent. and to leave less than .50 per cent. of sugar in the exhausted chips. We tried the use of lime in the cells. Practically, when making white sugar, we can not work the battery hot enough to obtain clean juice. We try to keep the battery at about 90° C.

Further experiments have also been made in the application of diffusion to sugar-cane by Prof. W. C. Stubbs at the Kenner Sugar Experiment Station. A full report of this work will be published in a forthcoming bulletin of that station. In the Louisiana Planter and Sugar

Manufacturer of December 1, 1888, a report is found on a part of the work done. As high as 240 pounds of sugar have been obtained per ton of cane. The results of the work are in every way encouraging.

From the above it is seen that diffusion with sugar-cane is an assured success, and we may expect to see it gradually displacing the milling process throughout the sugar-producing world.*

THE USE OF LIME IN THE DIFFUSION BATTERY.

The use of carbonate of lime in the diffusion battery and the patent obtained for this process by Prof. Magnus Swenson are fully discussed in Bulletin No. 17, p. 61, et seq.

Since the publication of that bulletin and of Bulletin No. 14, further experiments at Conway Springs have demonstrated that the method originally proposed by me for the use of lime to prevent inversion in the battery by evenly distributing finely-divided lime upon the fresh chips has proved satisfactory. An apparatus constructed by Mr. E. W. Deming succeeded fairly well in evenly distributing the lime over all the chips entering the cell in such a fine state of division as to prevent any portion of the contents of the cell from becoming alkaline. The lime was prepared by air slaking and sifting through a fine sieve into a barrel covered by a cloth to protect the laborer.

During the past year the use of lime in the diffusion battery for clarifying the juices has received a good deal of attention. The first person who proposed this process and took out a patent upon it was Mr. O. B. Jennings. Letters patent, No. 287544, dated October 30, 1883, were issued to Mr. Jennings on an application filed on the 2d of April, 1883. Following is an abstract of Mr. Jennings's patent:

Be it known that I, Orlando B. Jennings, of Honey Creek, in the county of Walworth and State of Wisconsin, have invented certain new and useful improvements in the manufacture of sugar from sugar-cane, sorghum, maize, and other plants, of which the following is a full, clear, and exact description:

This invention relates to the manufacture of sugar from different sugar-producing plants, including sugar-cane, maple, sorghum, and maize; but it has more especial reference to defecating the juice in the stalks of sugar-cane, sorghum, and maize, and extracting the juice from the residue or bagasse for subsequent boiling into sugar and sirup.

In making sugar from sugar-producing plants with my invention, it is my purpose to extract and utilize all of the saccharine juice and to obtain entire control of its defecation, so as to make a sirup free from foreign matter and elements of fermentation. By it the juice in evaporating is free from skimmings or precipitates, that are always liberated in the ordinary method of extracting, which waste my invention avoids.

Applied to the manufacture of sugar from cane and other stalks, the invention consists in a process of preparing said stalks for the more perfect extraction of the juice by reducing the same to a finely-comminuted or dust-like condition, and whereby the juice cells are thoroughly crushed and ruptured. This part of the invention also includes a combination of circular saws, forming a compound saw, for reducing the canes or stalks to such finely-comminuted condition, likewise sprinkling or mixing

---

*A report of the work done in Louisiana during the past season will soon be issued as Bulletin No. 21.

with said dust, before defecation, dry lime or lime whitewash in powder. Such lime combines with the acid in the dust, and upon a suitable application of heat to the whole forms double precipitates at one and the same time.

Furthermore, the invention consists in a process of precipitating the matter in the cane-juice cells and cane pulp, or in the juice of any sugar-producing plant, however obtained, by exposing the juice or material under treatment to a temperature of over 212° F., and subsequently removing the juice from the woody or precipitated matter by washing the same with currents of water. In carrying out this part of the invention I use a cylinder or other suitable vessel in which the temperature is raised to the required degree (about 212° F.) for defecation and precipitation of the matter capable of being precipitated, whether the same be contained in sugar-cane, sorghum, and maize stalks, reduced to dust or not, or in any saccharine juice, including maple sap, the temperature varying from 228° to 267° F., according to the ripeness of the material under treatment and other conditions. This vessel is suitably constructed or provided with means to admit of the introduction of the material to be treated; also, to provide for the forcing out of the exhausted bagasse or refuse, and for the introduction of steam while and after charging it; likewise, steam to act upon the condensed water and released juice and force them out through a filter. Means are also provided for running the wash-water from a series of tanks in succession through said vessel, to act upon the charge therein, and an arrangement of defecating-tank connections for introducing scum, sediment, and sweet wash-water upon a succeeding charge.

In the process of extracting the saccharine matter of cane, the mixing with the comminuted cane, before the passage of the same into the diffusing apparatus and the defecating of the same, of dry lime or lime whitewash, whereby the material will be thoroughly defecated without the liability of the admixture therewith of the precipitate of the lime, substantially as described.

The combination with the diffusing tank of one or more defecating tanks, to which the juice is delivered from the diffusing tank, and pipes provided with valves for drawing the skimmings, settlings, and sweet water from said defecating tank or tanks and passing the same into the diffusing tank or vessel, essentially as and for the purposes herein set forth.

In combination with the defecating tank, diffusing tank, and a suitable evaporator, the settling tank provided with a discharge pipe for running the juice into the evaporator, and with means for passing its sediment into the diffusing tank, substantially as described.

It is seen that Mr. Jennings makes a broad claim for the application of the process of clarification in the diffusion apparatus for all sugar-producing plants. Mr. Jennings has claimed that the process devised by the Department for the use of lime to prevent inversion in the battery is an infringement on his method. Any one who will carefully examine Mr. Jennings's claim, as set forth by himself in his application for a patent, will see that the two processes are entirely different, not only in principle, but in the method of application.

In a letter to the Rural World, published on the 13th of December, 1888, I endeavor to make this matter clear; following is a copy of the letter:

UNITED STATES DEPARTMENT OF AGRICULTURE,
DIVISION OF CHEMISTRY,
Washington, D. C., December 1, 1888.

EDITOR RURAL WORLD: I have read, in the Rural World of the 22d of November, the letter from O. B. Jennings, of Grover, Colo., in regard to his patent for clarifying cane juices in the diffusion battery.

Mr. Jennings is laboring under the mistake that I have been using his process and spending five years on what he showed me how to do at first. This is a complete misapprehension of the case. I have never denied to Mr. Jennings the honor of inventing the method of clarifying cane juices in the diffusion battery; in fact, long before his letter in your paper appeared I wrote a note to the New Orleans City Item, specifically claiming for him the honor of the invention which had been attributed to another source.

It is important to sugar-makers, either present or prospective, to know the following points, viz:

(1) The process of using carbonate of lime in the diffusion battery is a patented process which can only be used under royalty or by permission of the inventor, Professor Swenson.

(2) The process of clarifying the cane juices in the diffusion battery is a patented process and can only be employed under royalty or by permission of the inventor, Mr. O. B. Jennings, of Grover, Colo.

(3) The use of dry lime or lime in any form in the diffusion battery to prevent inversion is a process devised by the Department of Agriculture, and offered free to all sugar-growers in this country. Under proper chemical control it is more efficient than the use of carbonate of lime.

I will say further that I have never tried in any way to use Mr. Jennings's process, since in an ordinary diffusion battery it would be wholly impossible to do so. The high temperature which he requires for the proper clarification of the juices would render the circulation of the liquid in the battery almost impossible.

Respectfully,

H. W. WILEY,
Chemist.

The process of using lime in the diffusion battery for clarifying purposes it is claimed has been successfully practiced in Java and Australia.

Prof. W. C. Stubbs has also used it with success at the sugar experiment station at Kenner, La.

Col. E. H. Cunningham of Sartartia, Tex., has also used the process with success, as is indicated by the following letter from him, published in the Louisiana Planter of December 1, 1888:

My diffusion battery is now working nicely, and I am very much gratified at the results obtained. Diffusion is a success beyond a doubt. I am now working sugars by running the juice direct from the diffusion cells to the double effects without any clarification, except using a little lime in the diffusion cells.

I shall be glad to have a visit from you or any of your friends who feel an interest in diffusion.

The process of ordinary clarification, in my opinion, is more favorable to the production of a pure sugar than any form of clarification in the cells of the battery. The process as practiced at Kenner and Sugar Lands, however, differs from that described by Mr. Jennings in working at a lower temperature.

COMPARISON OF TOTAL SOLIDS DETERMINED BY SACCHAROMETER AND DIRECT DRYING.

During the season of 1887 I instructed the chemists at the Fort Scott station to make a series of comparisons between the total solids as determined by our standard saccharometer and by direct weighing.

The desiccations were to be made in flat dishes partly filled with loose asbestos or clean sand. The purity co-efficient of the juice as shown by the spindles appeared too low to permit so large a yield of dry sugar. As was expected, the total solids as determined by direct weighing were found considerably less than were indicated by the spindles. The ratio of each variation was not the same, but a large number of determinations established a mean rate of variation which will make it possible to approximately correct the reading of the common spindle. At Magnolia last year similar experiments were made with the juices of the sugar-cane, but these were not extensive enough to fix the rate of variation for those juices. Following is a record of some of the work done here:

Comparison of total solids.

No.	Total solids by spindle.	Total solids dried in dish.	Difference.	Total solids in hydrogen.	Difference.
	Per cent.	*Per cent.*		*Per cent.*	
6020	12.60	11.93	.67	....	....
6065	15.20	13.54	.66	....	....
6070	13.20	12.87	.33	....	....
6074	12.20	11 48	.72	10.94	1.26
6075	11.50	11.04	.46	10.84	.70
6076	13.30	12.85	.45	....	....
6079	12 30	11.77	.53	11.59	.71
C081	12.50	12.00	.50	11.65	.85
6083	16.30	16.04	.26	....	....

The determinations in hydrogen were made in a specially constructed apparatus, consisting of glass cylinder furnished with a glass stopper carrying two tubes with stop-cocks for displacing the air with an atmosphere of hydrogen. The juice was absorbed by a dried paper coil and supported in the cylinder on a disk of wire gauze resting on a lead tripod. The cylinder contained 25^{cc} of strong sulphuric acid. The cylinder carrying the coil was placed in a steam bath filled with dried hydrogen at 100°. The stop-cocks were then closed and the whole apparatus left at the temperature of the steam for five hours. The sulphuric acid absorbed all the moisture, and after cooling and filling the cylinder with dried air the coil was removed and weighed in a closed holder.

The determinations in flat dishes were made by drying 2.5 to 3 grams of the juice at 102° for five hours. Scarcely any difference was noticed between the results given by the plain dishes and those filled with sand or asbestos, except in the work at Conway Springs.

In the determinations made here in plain dishes the percentage of total solids was 4.68 per cent. less than by the spindle. In the determinations in hydrogen they were 6.94 per cent. less. The determinations in hydrogen, therefore, will show 2.26 per cent. less total solids, calculated on the number given by the spindle, than those obtained by drying.

At Douglass, Kans., the normal juice, calculated on the data furnished by the spindle, showed a loss of 8.61 per cent. in total solids when dried in open dishes.

At Conway Springs this loss in plain dishes was 7.24 per cent., and in asbestos 8.23 per cent.

With diffusion juices these losses were, for Douglass, 11.34 per cent., and for Conway Springs 9.67 per cent. in plain dishes, and 10.83 per cent. in asbestos.

The mean loss for normal juices at Douglass and Conway Springs was 8.36 per cent.

For the diffusion juices the mean loss was 10.61 per cent.

It appears therefore that a saccharometer of the standard Brix variety, as standardized by a pure cane sugar solution, must be corrected by fully 10 per cent. of its readings in order to give an approximately true indication of the total solids found in the diffusion juice of Kansas sorghum. For sorghum grown in New Jersey, which was the source of most of the juices examined here, the correction will be only about 7 per cent.

I am having constructed some saccharometers with scale to read as indicated by the above corrections.

The apparent purities of the sorghum juices will be considerably raised by this correction; thus at Douglass the purity of the normal juice is raised from 59.63 per cent. to 65.31 per cent., and at Conway Springs from 66.70 to 72.76 per cent. The purity of the diffusion juices of the two localities is raised from 58.59 to 66.86 per cent., and 62.92 to 71.13 per cent., respectively.

SUMMARY.

It has been my duty during the past few years to report the facts concerning the sorghum industry as they were developed by the researches of the Department and of others. These facts have been of a varied nature; sometimes they have been favorable to the industry and sometimes unfavorable, but in all cases they have been fully set forth and commented on in the light of knowledge at hand. In these investigations I have been unmoved by the abuse of interested parties, which I have received on account of my unwillingness to conceal the weak points of sorghum. It was thought when Bulletin No. 18 was issued that the experimental work on the part of the Department with sorghum was finished, and in that bulletin a summary was made of the investigations conducted in the United States during the past twenty-five years. In that bulletin I expressed the belief that with cane as rich as had been produced in Kansas on a large scale it was probable that a yield of from 80 to 90 pounds of sugar per ton of clean cane can be secured. The results of the past year confirm me in this opinion and indicate that, with wise management and careful control and proper selection of locality the sorghum-sugar industry may be made financially successful. In previous pages I have endeavored to set forth carefully some of the things which must be considered in order to secure the above result; but it must be remembered that my individual opinion is simply based upon the study of the facts which have been set forth. These data are accessible to every one who cares to make a careful study of the subject, and therefore each one interested has every opportunity to form his own opinion concerning the matter. Since it is my business to investigate rather than to theorize, I have contented myself chiefly with reporting facts rather than expounding theories.

REPORT OF H. A. HUGHES, RIO. GRANDE, N. J.

The whole season of this year has been devoted entirely to experimental work, with the object of securing additional light on crop growing, manufacturing, and commercial problems.

The past season was the end of a series of crop growing, covering a period of nine years, and fully confirms the fact that the safe time for planting Orange cane, after allowing for variations of climate, had passed.

The Amber cane had gone by its season by September 23, at which time the cutting had commenced, and the Kansas Orange had very little ripe seed on it; the Late Orange contained very little ripe seed, and a large number of the plumes did not even have seed formed in them.

The crop was all harvested by November 1.

The usual frosts and ice were met, with results described later on.

Analyses.

Description.	Sucrose, per cent.	Brix, per cent.	Purity.
Amber.......................	7.35	13.70	53.60
Kansas Orange..............	8.47	14.21	59.60
Late Orange.................	6.74	12.01	53.80

The Amber was used to break in the new machinery, not being considered worth working for sugar. The Kansas Orange was all worked for sugar and gave yields of fine quality of 86 to 90 per cent. test; without washing, of from 65 pounds to 39 pounds per ton of field cane. The limit of crystallization can be marked at 55 per cent. purity. Crystals can be formed below this degree, but they are difficult to separate in the centrifugals.

The Late Orange was mostly below the crystallization point, and although crystals were attempted by the sugar-maker in order to find out the limit at which graining takes place, and several pans were actually grained the grains were so small that conclusions were reached adverse to the boiling for sugar of such material. Two weeks of the season were spent in breaking in the evaporator, and one week in solving the problems and testing the result on the battery of chips of different sizes, best for diffusion, and the balance of the time in regular working.

A lot of Kansas Orange seed was selected and distributed among twenty different farmers, thus repeating the experiment described under season 1881, except that Kansas Orange of the finest quality was used instead of Amber. The result was high and low test canes and large and small tonnage.

It is but just to say that many of these farmers had no knowledge of cane raising and followed their own notions. Those who had knowledge of our work and some experience raised high-test canes and large tonnage.

This season completes the circle of observations and records of crops for nine years. The data can be summed up, which shows the action of fertilizers on large masses of cane as it has been received at the sugar-house, and the proper and safe dates for planting each variety are determined. This will explain and answer many of the criticisms which have been published from year to year by parties who only saw this work from one season's stand-point. The following deductions are made from the analysis of more than 38,000 tons of cane, and cover a period of nine years. This table will be found convenient for reference, under the heading of season 1880 to 1888, inclusive. It must be borne in mind that these facts will only strictly apply to this climate and this soil; but until it can be proved that they will not apply elsewhere it will serve as a guide, and should be interpreted by taking into consideration the fertilizers used, the variations of the seasons, and the nature of the plant. These conditions are fully described.

Summary of record for nine years.

Fertilizers.	Season.	Seed procured in—	Planting.		Harvest.	
			Commenced.	Ended.	Commenced.	Ended.
Complete fertilizers	1880	Minnesota ..	May 24	May 24 ...	Sept. 22	Oct. 13.
Unknown	1881	Rio Grande	Not known..	Not known	Not known .	Not known
Pacific guano	1882	...do	May 24	June 11...	Sept. 4	Nov. 4.
Yard manure and begasse	1883	...do	May 4	May 23 ...	Sept. 10	Nov. 14.
Large quantities of stable manure and light dressings of phosphoric acid.	1884	...do	Apr. 15	May 6 ...	Sept. 6	Nov. 11.
Compost in small quantities.	1885	...do	Apr. 14	May 4 ...	Sept. 2	Nov. 11.
Small quantities of compost and muriate of potash.	1886	....do	Apr. 10	May 30 ...	Sept. 22	Nov. 16.
Large quantities of compost and muriate of potash.	1887	....do	May 9	June 3 ..	Sept. 8	Nov. 22.
Complete fertilizers and muriate of potash.	1888	....do	May 18	June 10...	Sept. 23	Nov. 1.

*Summary of record for nine years—*Continued.

| Fertilizers. | Tonnage per acre. | Polariscope test. | | Variety. |
		At commencement of campaign.	At end of campaign.	
	Pounds.	°	°	
Complete fertilizers	6, 000	14	14	Amber.
Unknown....................	Not known	6–14	6–14	Do.
Pacific guano	14, 000	10. 35	10. 56	Amber and Late Orange.
Yard manure and bogasse........	16, 000	9. 70	9. 14	Do.
Large quantities of stable manure and light dressings of phosphoric acid.	(*)	10. 96	12. 00	Do.
Compost in small quantities	11, 000	5. 04	10. 00	Do.
Small quantities of compost and muriate of potash.	12, 000	6. 00	9. 45	Amber, Kansas Orange, and Late Orange.
Large quantities of compost and muriate of potash.	(†)	7. 94	9. 48	Do.
Complete fertilizers and muriate of potash.	18, 000	7. 35	6. 54	Do.

* 8,000 to 32,000 pounds.   † 16,000 to 44,000 pounds.

The planting commenced on May 24, in 1880, and was each year earlier until it reached April 10, 1886, from which time the season was made later, including the present year, this completing the circle.

Season of 1880.—Ripening of the cane was traced with the polariscope, and when 14 per cent. of sugar was reached cutting began; and during the short time required to harvest it, no damage was received from winds or frosts. The juice was reduced to semi sirup in an open evaporator, and three weeks later was shipped to Philadelphia and worked for sugar, marking firsts, seconds, and thirds.

The cane was planted in hills 4 feet apart, and sufficient plant food used. The impression made by this crop was that rich cane could easily be grown on poor land, and that with a little more fertilizing large crops could be made. It has since been found by long and costly experiment that all the conditions for Amber cane were most favorable, excepting that a large tonnage could only have been secured by proportionately fertilizing.

Season of 1881.—Farmers raised the entire crop. The acreage was not known. It was proved this year that with seed from the same lot some farmers grew cane 14 per cent. of sugar in the juice, while others grew it with only 6 per cent. Many conjectures were made, and the impression prevailed that some lands were suitable for cane and others unsuitable. It was, however, apparent that all who had the best reputations for farming raised the highest testing canes.

Season of 1882.—Cane was grown by the company. Pacific guano high in nitrogen was used, and only Amber cane was planted. The Late Orange cane was grown only in sufficient quantity to supply seed for the next year. The nitrogen had the effect to keep the cane's leaves green for a long time, and even after frosts the cane remained in good condition, and was on November 4 higher in sugar than on September 4. Since we have had less nitrogenous fertilizing and more of other plant food this variety has steadily fallen in test, and the period during which it retains its highest sugar content has been shortened.

It is not safe to depend on this variety of cane for the whole season, even if nitrogen is used largely with other plant food, because of its tendency to lodge and break with high winds.

Season of 1883.—Yard composts and begasse were used in such small quantities that the nitrogen did not stand out prominently. The Amber had gone by its season before October 8, and had not the Late Orange been substituted this season for sugar making would have ended on that day, instead of November 14, when the crop was all in.

Season of 1884.—Stable manure in large quantities, also a dressing of dissolved bone-ash from South America, rich only in phosphoric acid, was used.

The phosphoric acid ripened the cane fully two weeks earlier than usual, and although the leaves were dry the Amber cane held its sugar content without loss until worked up on October 11. The Late Orange was affected in the same manner according to its season, and although apparently dried up, too, still held its sugar. Mill juice tanks containing 6,000 gallons were quite common, testing 13 to 13½ per cent. of cane sugar from October 11 to October 29, after which time there was a gradual falling off until November 11, when the tanks stood 12 per cent. and 77 purity. This ended this season, as the crop was worked up.

The small experimental plots conducted by the State Experiment Station have always showed that by doubling the dose of phosphoric acid the cane sugar falls off seriously; but as it is my intention to deal only with cane in immense masses as found at the sugar house, I merely call attention to this fact.

This year produced nearly 400,000 pounds of merchantable sugar, and there was found by adding the sugar in the molasses, and the loss in the begasse as it came from the mill, that over 1,500,000 pounds of sugar were in the crop.

Molasses only was made from the begasse this season, diffusion being for the first time applied.

Season of 1885.—No phosphates were used and there was not enough compost to properly furnish nitrogen to the crop; still the nitrogen was felt, and when the season commenced on September 2, the cane was so green we at one time thought it would be better to stop work. When work was begun, the Amber cane contained 5.04 per cent. of cane sugar and increased to 8.8 per cent. on September 29, when the variety was all brought in. The Late Orange cane contained 10 per cent. of sugar when first cut, and gradually raised to 12.57 per cent., slowly declining to 10 per cent. by November 11, the end of the season. This crop was planted practically at the same time as the crop of 1884, and harvested at the same time. Had a large quantity of nitrogenous fertilizing been used the sugar contents would have been much higher. Small quantities of nitrogen on lands deficient in organic matter will make poor crops.

This was our experience again and again, and to secure immense crops high in sugar, potash should be combined with nitrogen.

Season of 1886.—Small quantities of nitrogenous fertilizer and light dressings of muriate of potash were used. The crop suffered severely for lack of food. During the season, where plenty of nourishment had been supplied, the crop came to the standard. When this was not the case, the Amber seed remained in a milky state for a long time and soured as it stood in the field, after three days of abnormally hot weather, making the cane unfit for sugar making. The Late Orange suffered from lack of nitrogenous fertilizing and the sugar test rose and fell in proportion as this food and potash were present; but being a longer feeder it did not suffer throughout the season so much as the Amber.

The Kansas Orange was introduced this year and, being a stranger, the ground was properly selected, and composts and potash applied in sufficient quantities, a 12 per cent. cane with purities over 70° being its record. The record of the Late Orange cane, for the balance of the season, is high and low test, according to the land; finally ending, with the crop all harvested, with a test of 9.45 per cent. This crop discouraged the sugar company notwithstanding the gains by diffusion, which process had been introduced in 1884. Local agriculturists pronounced the verdict that the lands being exhausted by continual cropping were ruined and unfit for crop of any kind. The plantation was then sown in clover; no fertilizing was done. The farmers laughed at the notion that land unable to grow large cane crops could be expected to grow grass, but it did ; and the clover crops on these lands have been unprecedented and are the envy and wonder of local farmers, and judging the land from the farmers' own stand-point, it is to day in better condition than ever before. The clover had found the missing nitrogen and furnished organic matter.

A lot of land on these farms grew poor cane for years, and in 1887, instead of planting it with clover, composts and potash were supplied and cane planted ; by planting the ground with twice the number of hills to the acre, portions of the land approximated 28 tons of cane to the acre.

Season of 1887.—The cane was planted from May 9 to June 3, and the late varieties failed to mature properly. A good dressing of begasse yard compost, and potash was used. The crop was doubled by planting 3 feet by 24 inches; purity ran about 64° and tests were good. The Late Orange cane ripened sufficiently to retain its sugar in crystallizing quantities through frost and ice, until December 5. Particulars of this season can be found in Bulletins Nos. 17 and 18 of the Agricultural Department, and in reports of the New Jersey Experiment Station. A small plot was fertilized with large quantities of nitrogenous manure and planted with Amber seed grown in 1886, from which no cane sugar could be made. The cane was tested on September 7, 1887, and was found to test 13.35 per cent. cane sugar; brix, 17.21°; purity, 78°; and it remained a long time after in fine condition. The same day milled chips from a field planted from the same lot of seed and fertilized with potash and phosphoric acid, polarized 8.88°, and had a purity of 63 61.

Season of 1888.—Only complete fertilizers were used on one field, and muriate of potash was spread on another field that was poor and had never been in cane. The hills were 3 feet by 24 inches. Amber cane was planted on May 18, and Kansas Orange and Late Orange from May 19 to June 10. A cold, wet June followed, and the result was unripe cane. The crop was taken off between September 23 and November 1. The Amber cane was very poor in sugar. The Kansas Orange ran from 9.58° to 8.25°. The stand on one field of Orange (Kansas) was preserved intact from cut and wire worms, by patches of volunteer canes, where seed had been stacked previously, and some seed had been left on the ground. The worms gathered where plants were the thickest, leaving the hills almost unmolested. When the ravages are feared seed could be sprinkled down the center of the rows, and afterwards be destroyed by the cultivator without extra expense. They only destroy while the plants are very small and disappear with the return of dry, hot weather.

The Late Orange tested from 6.94 to 6.54. Scarcely any seed on this variety was ripe, and in a great many of the plumes seed was not formed neither had the cane power to resist ice and frost. These facts prove conclusively that the safe time for planting Late Orange has been passed. It is possibly true this variety might have been very rich in sugar, with a late fall and hot weather during June and September; but this risk is not a safe one, and as it positively can be avoided by earlier planting it should be done.

OBSERVATIONS.

The time for planting cane in this climate is, for Early Amber not later than May 20; Kansas Orange, not later than May 10; Late Orange, not later than May 1. Ten days earlier can safely be risked.

Nitrogen prolongs the vitality in cane.

Nitrogenous fertilizers combined with potash is the best combination for large crops and high testing juice. Phosphoric acid hastens the ripening of the cane about two weeks, and too much phosphoric acid reduces the quantity of sugar in the juice.

Potash makes large and strong stalks. If canes are desired to be worked after frost and ice, they must be supplied with ample food, be well grown, and of a late variety. If canes are not well advanced when frosts and ice strike them, they will not be able to hold the cane sugar long.

The earlier the variety the later it should be planted. If canes increase rapidly in cane sugar soon after frost strikes them they will soon be worthless for sugar-making. If they do not increase at all, or very little, they will remain good for a long time, providing the frost was severe enough to kill, or almost kill, the leaves. The Amber has less power to resist frost and ice than Kansas Orange, and the Kansas Orange less than the Late Orange. The time which the sugar remains in high percentage in the cane is largely under the control of the cultivator. In all attempts to improve the seed by selection and increase the sugar and purity, the cultivation must be taken into consideration. High testing seed will make poor testing canes, if plant food is not present in sufficient quantities, or if the cultivation is neglected. Poor testing seed will give high testing canes if the seed is of a good variety, and ample food has been supplied, with good cultivation.

Canes can not be grown, rich in sugar, by starving them. Ground well supplied with plant food and badly cultivated will give very small canes, rich in sugar. That there are other peculiarities in other varieties is shown plainly in the case of the White African. Although planted late last spring, and the ground fertilized precisely like the Amber and Kansas Orange, it contained this year 12.30 per cent. cane sugar, purity 69° on September 27, time the field was cut. The seed was given to the writer by Dr. Collier along with sixty-eight other varieties in 1883, all of which were planted; but for certain good reasons this cane was the only one selected from the lot. It has been grown since then each year, always giving high percentages of sugar. Some of its peculiarities are, viz, the unusual toughness of its stalk, when overripe, and its great strength at all times.

It is hard, for some unexplained reason, to get a good stand. The seed is white, and local millers, with their crude appliances, have told me that they can get 30 pounds of flour from 1 bushel of seed, which, mixed with a small proportion of wheat flour, is preferred to buckwheat. The birds ravage the seed, and will select it from a hill planted with mixed Orange and Amber canes, leaving the other varieties unmolested. In order to be protected from these depredators and secure the seed, plots of sufficient size must be raised and calculations made for this loss. It has been found true here that they will not take quite all the seed from 1 acre in a season, consequently plots of 5 or 10 acres are comparatively protected.

The purity of the canes of this variety has been noticed as high as 77.92°.

The cane has not been properly studied, and the birds have taken nearly all the good seed from the acre raised this season.

MANUFACTURING.

I will confine myself, in my report, to methods adopted for the first time this year.

Sawdust filters.—It has always been found that filtration of the juice through some medium that would remove the particles of matter mechanically suspended was necessary. For two years, filter presses were used. It was found if the juice was acid they soon became gummy and refused to run; if the juice was alkaline it would filter much better, but gave highly colored products.

Last year Dr. Wiley advised the use of sand. This gave good results for a time, but gradually ran slow and failed to give satisfaction. The size of the filters, in proportion to the juice worked, was very large, and it soured easily.

During the past winter experiments were constantly carried on with the hope that something practical, cheap, and easily handled would be discovered. Experiments were made with bone black, coal, sand, gravel, oat straw, wheat straw, grasses, sedges, excelsior packing, and many other things, all of which proved unable to do the work required, were too costly, bulky, or in some other way not desirable. It was accidentally found that the coarse sawdust as it came from the mill would do the work.

Shallow filters are better than deep ones, and in well-conducted experiments the juice was so well cleared of its mechanical impurities that it appeared to be bleached.

Examinations of the filters showed, among other things, soot from the chimney, mud, and dirt. The juice was actually cleansed. The filter used in this season's work was constructed as follows: A board twelve inches wide was cut in four pieces and a box made 4 feet long by 2 feet wide; a wire screen with one-sixteenth of an inch mesh was fastened on the bottom, and three inches of sawdust placed within it. Care should be taken that something should be placed over the sawdust to break the fall of the juice and prevent guttering.

It was found in practice that 1 bushel of sawdust was sufficient to filter the juice from 15 tons of cane, and that the filter should be renewed every twelve hours.

It may also be well to state that the hot juice as it came from the evaporator was run through a sawdust filter, removing scum, scale, dirt, etc.

Double shredding.—In 1885 samples were taken of the exhausted chips as they came from the German diffusion battery and it was found that better diffusion had taken place in small chips than from a larger size; and last season this was found true also of the battery which was then being tried for the first time. All attempts to obtain a chip of the size required failed, owing to the following facts: If the knives of the shredder and the cutting bar were placed so closely together that the small chips might be made either the shredder would not feed fast enough or the knives would clog with the fine cane and stop cutting. It was found this season that by making the ordinary cut first and afterwards allowing the edge of the knives to project beyond the cylinder very slightly, and by moving the cutting-bar closer and passing the previously cut cane through a second time, the chips could be made as fine as possible or as desirable.

It was found in actual work that baskets of cane filled with chips of the customary size weighed 160 pounds, and packed in the same way with the re-shredded chips weighed 212 pounds, thus increasing the capacity of the battery, and by its close packing increasing the density of the juice.

It is to be hoped, notwithstanding the brilliancy of these results, that manufacturers will not at once attempt to double shred their chips, because the second time they go through they are not self-feeding, and machines should be invented and proven equal to their task before a commercial season should be risked.

Evaporator.—In accordance with your instructions, I constructed an open evaporator to be run by crude oil (petroleum). Parallel brick walls 13 inches thick, 34 feet long, and 24 inches high were constructed. At one end was an iron stack, and at the opposite end were the burners. Upon the walls was placed an open evaporator of sheet-iron 1 foot high, 30 feet long, and 4 feet broad, divided by partitions 8 inches apart, 6 inches high, and 45 inches long. The juice entered the pan over the burners, discharged at the opposite end, traversing a distance of about 164 feet in twelve minutes. The skimmings remained at the end over the burners and were easily removed. As this was the first time, to my knowledge, that crude oil had been applied to sugar work, I was able to collect little data to guide me. After examining personally the burners in use for steam-boilers, I finally adopted one belonging to H. W. Whiting, of Philadelphia. He advised me to place three burners at the end, and inserting in the brick-work, at intervals of 1 foot, inch pipes, to extend completely through the walls and flues and to be perforated with holes one-fourth of an inch in diameter and 3 inches apart. The intention was that air should pass through the end of these pipes, then through the perforated holes into the flue, and thus aid combustion.

The burners were made from 2-inch pipes with a T fitting opening at the bottom to supply air on the Bunsen burner principle; the oil passed through a quarter-inch pipe, through a cock into a 1¼ inch coil 1¼ inch in diameter, so placed as to receive a large portion of the heat from the burners; there is also a quarter-inch steam pipe leading into the end of the pipe, so that the oil and steam can be mixed as it passes into the hot coil, or superheater, as it is named. When the oil is converted into gas from the superheater it passes into the Bunsen burner and is forced through it by another steam jet and burned from the opening.

In our first experiment Bradford crude oil was used, and in our final experiments black residuum of the refineries, which I have been informed is the product left behind after the light oils have been distilled off.

In practice we could find very little difference in the heating of the two oils. Lima oil could not be had in quantity less than 6,000 gallons; consequently it was not used.

It was found in starting the burners that a stack 10 inches in diameter was too small, the effect in practice being to cause explosion of gas.

A stack of 24 inches diameter was substituted; this stopped all explosions, but wasted the heat. Dampers made of fire-clay were then used, and it was found that after the superheater was hot enough to generate gas freely the dampers could be safely closed. Care had been taken in constructing the dampers to arrange them so that there was left on the sides a space equal to about 12 inches square after they were in. A further improvement in the heating was made by filling in next to the stack with dirt. This bank of earth was then extended back into the flue for about its length and paved on the top with bricks. There was left a space of about 9 inches between the pavement and the bottom of the evaporator; and in filling in the flue the combustion pipes were covered up for the length of the embankment. The combustion pipes directly in front of the flame were soon burnt out. No detrimental effects being perceptible from the loss of this air, it is safe to conclude that they were of no value.

The owner of the burners thought we would evaporate at least 15 pounds of water for each pound of oil burned, and hoped we would reach 18 or 20 pounds. The record of the best day's work shows $7\frac{8}{10}$ pounds. It is but just to say that the evaporator was entirely too large for the work it had to do, and the walls had time to cool before starting each day. Now it is found that if the walls and surrounding mediums are much lower than the temperature of the gaseous product of the Bunsen burners, condensation takes place and the oil is fried, as it is called, instead of being generated into gas, which is wasteful in the extreme. One-third of all the oil burned was generally used in starting the burners each day. Another source of loss long evaded our researches. It was caused by using cocks to feed oil to the superheater. A common quarter-inch globe valve was substituted for the cock, which brought the burners under full control and enabled us to burn only one-quarter as much oil. I make the suggestion that pipes for supplying oil to the superheater should should be less than one-quarter inch; that globe valves less than one-quarter inch be used, and that threads that regulate these valves be made as fine as possible, so that they may have the most delicate adjustment. I can not tell the saving of all these apparent improvements, because, I had not time to get the record properly. Taking the record as it is and counting the price of oil at $1.25 per barrel, about one-half of the water was removed from the diffusion juice of each ton of field cane for 31 cents per ton.

The advantages of the evaporation are: (1) Cleanliness and freedom from smoke and ashes; (2) the little attention required to run it; (3) the good and rapid work done.

With rapid running the inversion is almost nothing; in fact, after evaporation it is sometimes higher in purity than before, after removing the scum.

It should be remembered that the unrefinable Lima oil has been quoted at the wells for 15 cents, which would lower the price for evaporation of their juice in that section to 4 cents per ton. The loss in starting could be avoided very much by proportioning the evaporator to the size of the house.

The Battery.—The designing, building, and breaking in of such an apparatus as a new diffusion battery on an entirely new principle could not but prove a gigantic task.

The object of the battery at first was to make a cheap diffusion battery, applicable to small houses; second to make thick juice.

For three seasons laboratory experiments were carried on at Rio Grande and dense juices made by diffusion, equal to mill juice from unstripped cane, and the principles by which this juice was obtained were incorporated in this battery.

The season last year was devoted completed to the breaking in and finding out the rules governing this machine.

The ram constructed to lift the baskets, last season, worked slowly. When making some changes this fall the cause was located and corrected. Owing to this mechanical difficulty and being forced to take off a crop promptly, it was not until later in the season that plans could be put in practice which would remedy defects in heating and extraction. This was tried with temporary arrangements, but the results were considered so high that it was objected to on the ground that the time during which the experiments were conducted was too short to thoroughly demonstrate the facts.

The chemist of the New Jersey Experiment Station, after carefully going over his work, says, reporting on this experiment:

The best work accomplished by the Rio Grande battery was 90 per cent. extraction, dilution 11.5°; purity, declined 1ᶜ.

The cell necessary for heating the chips properly and thickening the juice is placed outside of the battery and is called the eleventh cell. This year this apparatus was added to the regular work, and from the first day never failed to give satisfaction. It is found that when the cane is carefully packed into the baskets the gain is not so great as when the baskets are loosely packed; at such times the full value of the eleventh cell appears, gaining 2° to 3° Brix.

The entire apparatus worked without delay, and the mechanical arrangements were very complete. For a battery of 40 tons, the baskets and cane together will not weigh 400 pounds, and the lift will be considerably less than 4 feet; consequently $400 \times 10 = 4,000$ pounds to be lifted, and $4,000 \times 4 = 16,000$ pounds to be raised 1 foot high at each movement of crane. The crane makes twenty movements in an hour or once every three minutes; consequently $16,000 \div 3 = 5,333$ pounds raised 1 foot high each minute, or less than one-sixth of a horse-power is required.

There is to be added to this the cost of raising the water for supplying the battery and the movement of the juice; but with these all added the cost for power is found to be merely nominal.

With double shredded cane and actual running, the dilution was reduced to 4½ per cent. and approximated the mill juice within four-tenths of a Brix, with a loss of only 16 pounds of sugar left in each ton of cane. The Brix of the milled diffusion chips showed from 1½ to 2. Without double shredding the battery gave within ½ to 2° Brix of the mill juice, and left about 16 pounds of sugar in the chips per ton of cane. The purity fell off one to two degrees, but it must be remembered that no chemicals were used to prevent it. There is always a percentage, about 2 per cent., of leaves and sheaths which pass the cleaners, and as their purity is very low they must reduce the purity of the diffusion juice. Lime and its salts and sulphites have been used in batteries, and have appeared to give juice of as high a purity as the mill juice; but it would have to be shown that some of the glucose had not been destroyed before the point can be positively settled. Besides, alkalies used on the fiber in the cells and clarifiers where the fiber is present are believed to produce gum.

It has been observed this season that when scum raised in the chips from heat, while diffusion was going on, that the juice coming from this battery was higher in purity than mill juice. There is no evidence that the air passes through the cane, while being diffused, except when first heated; neither do the juice or chips turn black while diffusing, as is supposed by some; and the color of the juice will compare favorably with the mill juice.

INVERSION AND CLARIFYING.

Considerable inversion has taken place in the house this season. The most of the inversion takes place by permitting the juice to stand hot for a considerable time in tanks, and in process of manufacture this should be carefully avoided.

This is the third year during which we have used no clarifiers, and the writer does not see what use they are with the present knowledge of the juice. Alkalies used too freely in the battery or in the clarifiers when fiber is mechanically suspended are thought to produce gum and prevent crystallization, although the instruments may show no loss from inversion. After the juice has been filtered, the addition of alkali in not too large quantities, so that the juice would be neutral, or, better still, slightly acid, would no doubt prevent some inversion. The correct method of properly clarifying the juice of the sorghum so that the " not sugar" parts can be precipitated, and the purity be made to gain largely, is not known to the writer. Rapid running in the diffusion battery and quick running in the open evaporator will almost entirely prevent the inversion of sugar.

COMMERCIAL POINTS AND AUXILIARY HOUSES.

The auxiliary houses have been steadily kept in view during the season's work, and the fact has been remembered that the industry will spread and succeed at a much quicker rate if the capital necessary to conduct the business is kept as low as possibly consistent with good management. The cost of building sugar-houses is reduced to a minimum, and labor saved. There is no good reason to expect to make money out of the sorghum business unless conducted on sound business principles. The knowledge of the business is now advanced to such a point that there is nothing to prevent accurate calculations being made. The cost of the machinery, the work it can do, the labor required to run it, the cost of the cane, the yield and quality of the product can now all be closely estimated.

Sugar-houses built without definite ideas of the work to be done or machinery added piece by piece, without plans or contracts, and such machinery as clarifiers, as filter presses, and bone-black drones added, with the expectation of only making white granulated sugar directly from the juice, will be certain to bring financial failure and disappointment to its projectors, unless the capital is heavy enough to stand the strain, or the parties are willing to make experimental work of their plants and pay the price for doing it. Notwithstanding the closeness with which all these calculations can now be made, the following should be remembered. I have never known a sugar-house of any kind to be made so complete and be in such fine running order that it could be depended on to make a commercial success the first season. Either its water arrangements will fall short of expectations, or the boilers fail to be large enough, or strikes and delays will detain the machinery, or castings will be broken in shipping, or some minor points will be badly proportioned or too weak, foundations will prove not sufficiently secure, shafts will be found out of line, etc. All this will occur, not from any bad management, but because the nature of the work is such that the factory can only perform its task satisfactorily after being broken in on cane. The cane alone can give the necessary adjustment. Erroneous and disappointing calculations have been made by celebrated sugar engineers, in making calculations for sorghum, by using well-known standard rules for the evaporation of water as a basis for calculation; and repeatedly has machinery proved suitable for southern cane failed when applied to this work. The moral of all this is that in constructing new works there should be only enough cane raised the first season to break in and test the sugar-house thoroughly in every part, in order that when the machinery is called upon the succeeding season it would fulfill the work it had been calculated to do, without delay or hindrance.

The expense of doing all this should be allowed for in the capital account.

In some sorghum-houses, calculated to work 100 tons of cane a day, will be found strike vacuum pans of such large size that the cost of erecting them and the pumps necessary for their use, the large pipe fittings, and other paraphernalia will cost as much alone as would suffice to build an economical sugar-house of good size.

Experience has taught us that there is a limit to the size of sugar-houses, and that it costs very little more to man a 40-ton house than a 20 ton, and the proportionate cost of constructing is greatly in favor of the 40-ton plant. For sugar-houses of larger size, I can not yet give accurate data with safety.

The following is a plan based on calculations made from actual work already done; the rules known to govern the situation are carefully applied and full allowance made for such errors.

(1) The plan of a sugar-house complete for making sugar, according to the process in use at Rio Grande. The sugar will be brown or yellow, and test 86 to 90°. It is suitable for some domestic purposes and for refining. The molasses will be of fair color, suitable for mixing and baking purposes.

These sugars can be washed in centrifugals and made quite white, of high test, but at the expense of the yield. The proper place for them is in a sugar refinery to be remelted and run through black.

(2) An auxiliary house for making sirup and retaining the sugar in the sirup. Inversion would have to be as carefully avoided as possible. These goods or products would be very fine, and could be sold on their merits for immediate consumption, or find a market on their tests and color at the Central Sugar-House.

The large vacuum pans referred to are well calculated to work up goods in this condition, in immense quantities; the sugar could be remelted and run through black.

Dr. A. T. Neale, of the New Jersey Experiment Station, spent the season at the sugar-house. He had control of the chemical department, and results of his work will be found in a bulletin to be soon published by that station.

I respectfully submit the above report, with thanks to you personally for your uniform courtesy and support.

RECORD OF THE ANALYSES MADE AT RIO GRANDE DURING THE SEASON OF 1888.

By Dr. Arthur T. Neale.*

In addition to studying the construction, the arrrangement, and the management of the machinery, the chemist of this station attempted to determine, at least once each day, the percentage of sugar in the sorghum, as well as the percentage of sugar in the products from each piece of apparatus used in this house. Breaks occur in this record whenever it was necessary for him either to return for a day to New Brunswick or to devote his entire attention to some one point of special interest.

The house was not open for work until the 26th of September, and a few of the samples of cane analyzed about the 20th of that month were taken from the crop standing in the fields. Such samples were stripped and topped by hand. All of the other samples were drawn from cane which was cleaned by machinery. They represent in each case, approximately, 1,000 pounds of well-mixed shreds.

The varieties of sorghum planted were: Early Amber on field No. 1; Late Orange on that portion of field No. 12 which was worked after the 23d of October; White African on a portion of field No. 2 harvested on the 27th instant, and Kansas Orange in all other cases.

The exhausted chips were sampled as fast as they were removed from the battery; a roughly measured quantity being taken in each case from each one of ten baskets. These portions were subsequently mixed, subsampled, and milled in the usual manner.

With a few exceptions, the samples of diffusion juice were, in all cases, drawn from a tank holding 300 gallons. The samples of the evaporator product were also drawn from a similar tank. The record in detail is shown on the following page.

The averages drawn in this table prove that the cane crop in 1888, relative to that of 1887, was poorer in sugar by 0.75 per cent. and lower

* Bull. No. 51. New Jersey Agricultural Experiment Station, pp. 12-15.

in purity by 5.6 degrees. The farmers' explanations for this are: first, late planting; second, early frosts. In some cases the seed were dropped after June 1st, and in all cases the leaves were killed by the frosts which occurred this year on the 4th of October, or ten days earlier than usual. Late orange sorghum, in particular, seems to have suffered by these conditions, for while the cane was very large and apparently well developed, its juice averaged less than 6.5 per cent. of sugar. Its seed crop was practically worthless, for a very small proportion of tops had matured. In 1887 this variety was well developed when the first frost killed the cane leaves. Its juice then contained, approximately, 10 per cent. of sugar.

A comparison of the analyses credited to the cane and to the diffusion juices leads to the following calculations: One hundred pounds of solid matter, *i. e.*, sugar, etc., existed on the average in 715 pounds of cane juice, or in 920 pounds of diffusion juice; that is, cane juice was diluted 28.6 per cent. by the diffusion process. If a similar calculation is made from the records for the season of 1887, the dilution will be fixed at 25.4 per cent. The decreased purity of the diffusion juice was, each year, identical; it amounted to 2.1 degrees.

The exhausted chips, or diffusion bagasse, which represented 1 ton of field sorghum, contained on the average, in 1887, $40\frac{3}{10}$ pounds of sugar, or 35 per cent. of the total amount present in the cane. In 1888 the losses of sugar in exhausted chips amounted to $22\frac{4}{10}$ pounds, or 21 per cent. of the total amount present in the average cane for that year.

In 1887 the diffusion juice was concentrated in an open evaporator with the aid of steam ; it was reduced by this treatment to a fraction more than one-third of its original volume, at an expense of 4.3 degrees of purity, which was probably due to inversion of its sugar by heat. In 1888 the flame from burning fuel oil came in contact with the bottom of the evaporator; the diffusion juice passed in an unbroken stream over this heated surface, and was thereby reduced to less than one-half of its original volume. Its purity was decreased on the average by less than 1 degree.

The following will serve as a summary : In 1887, 65 per cent., in 1888, 79 per cent., of the total sugar in the cane was extracted. In this respect, therefore, the improvement has been very great. The diffusion process, in 1887, diluted cane juice by 25.4 per cent.; in 1888 this dilution amounted to 28.6 per cent.

The purity of the cane juice was influenced each year in the same manner and to the same extent, viz: decreased by 2.1 degrees. The concentration of the diffusion juice was accomplished in 1888 with considerable less than the usual losses by inversion.

Table of analyses at Rio Grande, season of 1888.

No. of field	1888.	Fresh chips.			Diffusion Juice.			Evaporated product.			Exhausted chips.		
		Brix corrected.	Per cent of sugar.	Purity.	Brix corrected.	Per cent of sugar.	Purity.	Brix corrected.	Per cent of sugar.	Purity.	Brix corrected.	Per cent of sugar.	Purity.
1	Sept. 20	12.90	6.62	51.1									
1	Sept. 26	13.38	6.84	51.1	11.50	5.88	51.1	29.20	15.13	51.8			
1	Sept. 27	13.70	7.35	53.6	11.61	6.02	51.9				2.27	1.18	52.0
2	Sept. 21	18.55	12.53	67.5									
2	Sept. 27	17.81	12.30	69.0									
2	Sept. 28	15.40	9.58	62.2	12.15	7.23	59.5	29.60	17.78	60.1	2.56	1.50	60.1
2	...do				11.95	7.23	60.5						
2	Oct. 2	16.00	9.56	59.4	12.37	7.19	58.1	23.45	13.75	58.6			
2	...do				11.69	7.01	59.8						
3	Sept. 22	14.75	8.92	60.5									
3	Oct. 2	14.10	8.38	59.1	10.60	5.98	56.4	25.88	15.01	58.0	2.09	1.21	37.9
3	Oct. 3	14.00	8.62	61.6	10.00	5.70	57.0	20.04	11.20	55.8	2.21	1.38	62.1
3	...do	13.43	7.90	58.8	10.90	5.99	55.0	20.00	11.06	55.3	2.70	1.56	57.7
4	Oct. 4	14.03	7.89	56.1	11.21	6.48	57.8	28.28	15.80	55.1	3.40	1.60	48.8
5	...do	15.37	8.72	56.7				24.85	13.58	54.7			
5	...do	15.57	9.06	58.8	12.30	7.45	60.5				3.56	2.07	58.1
6	Oct. 8	14.46	8.66	59.8	11.25	6.58	58.6	26.80	15.24	57.0	2.00	1.29	49.6
6	...do	14.46	8.37	57.8	10.60	6.10	58.5	22.40	13.09	58.4			
6	...do							22.66	13.13	57.9			
7	...do	14.79	9.19	62.1	9.92	5.96	60.1	20.44	11.52	56.4	3.60	1.96	54.4
8	Oct. 10							22.87	12.35	54.9			
8	...do							22.66	12.36	54.6			
8	...do	14.70	8.10	55.1	11.82	6.61	55.9	20.51	11.13	54.3	3.82	1.96	51.6
9	Oct. 11	12.64	7.48	59.1	9.58	5.61	58.5	22.43	12.33	55.0			
9	Oct. 12	12.60	7.91	61.8				22.26	12.30	55.2	2.80	1.76	61.9
10	...do	13.40	8.50	63.4	9.80	5.90	60.8	22.78	12.93	56.8	2.20	1.22	55.0
11	Oct. 18	14.00	8.37	59.8	8.90	5.23	58.7	25.06	13.05	54.5	2.11	1.24	58.7
11	. do	13.50	8.25	61.1	10.30	6.15	59.7				2.00	1.14	57.0
12	Oct. 16	12.85	7.68	58.4	9.88	5.51	55.7	22.69	12.60	55.5	2.56	1.39	54.3
12	Oct. 17	13.06	7.91	60.6	8.80	5.03	57.2	24.18	12.76	52.8	1.97	1.05	53.3
12	Oct. 22	12.43	7.69	61.9	11.90	5.95	57.7				1.54	0.92	60.0
12	Oct. 23	12.60	7.23	57.0	10.21	5.65	55.3				2.16	1.20	55.6
12	Oct. 24	11.42	5.78	50.6	10.27	5.06	49.2				2.56	1.20	46.5
12	Oct. 25	12.67	6.94	54.8	12.32	6.70	54.4				3.49	1.58	45.2
12	Oct. 26	12.76	6.63	52.0	11.96	6.21	51.9				2.77	0.91	32.8
12	Oct. 27	12.36	6.54	52.9	11.83	6.02	50.9				1.96	0.82	41.8
12	Oct. 29				8.88	4.68	52.7	17.77	9.58	53.8			
12	Oct. 30				10.61	5.59	52.7	24.80	12.49	50.3			
	Averages, 1888	13.99	8.23	58.5	10.87	6.10	56.4	23.55	13.06	55.5	2.58	1.37	53.3
	*Averages, 1887	14.02	8.98	64.1	11.18	6.93	62.0	32.40	18.08	57.7	4.03	2.46	61.0

* See Bulletin 18, p. 20, United States Department of Agriculture.

REPORT OF PROF. W. C. STUBBS, KENNER, LA.

LOUISIANA SUGAR EXPERIMENT STATION.

On April 6, 1888, two plats, Nos. 9 and 10, at the sugar experiment station were planted in sorghum.

PREVIOUS CULTURE.

No. 9 had been continuously in sorghum since 1886, and No. 10 in corn.

PREPARATION OF LAND.

The land was broken in the spring with four-horse plows, thrown into beds 5 feet apart, and seed sown and lightly harrowed in. Only a partial stand was secured, germination being prevented by a prevailing drought. It was thinned, wherever thick enough, to three stalks to the running foot. The cultivation consisted of off-bearing with two-horse plow, a hoeing, and returning the dirt with two-horse plow, and breaking out the middles with a large one and three-quarter Avery Advance double mold-board plow.

The excessive rains began in May and lasted till the middle of July, and prevented further cultivation. The varieties planted on these plats were:

1. Honduras seed, grown at the station.
2. Honduras seed, grown on the Teche.
3. Link's Hybrid seed, grown in Kansas.
4. White Mammoth seed, grown at the station.
5. White India seed, grown in Kansas.
6. Enyama, grown by J. P. Baldwin, of the Teche.
7. Early Orange, grown in Kansas.
8. Kansas Orange, grown in Kansas.
9. New Orange, grown in Kansas.
10. Golden Rod, grown in Kansas.
11. Honey Drip, grown in Kansas.
12. Texas Honey Drip, seed bought of Gumbrell, Reynolds & Allen, Kansas City, Mo.
13. Planted with seed from Department of Agriculture, but none came up.
14. White Minnesota Amber seed, grown in Nebraska.
15. Early Amber seed, grown in Kansas.
16. Early Amber seed, furnished by Department of Agriculture.
17. Kansas Orange seed from Kansas.
18. Link's Hybrid seed, grown at the station.
19. Early Orange seed, grown at the station.

Several of the above varieties were sent to the State Experiment Station, Baton Rouge, La., and to North Louisiana Experiment Station, Calhoun, La., and experimental plats planted at each station.

The varieties planted at Baton Rouge were Early Amber, Early Orange, Link's Hybrid, and Honduras.

They were planted in rows 4 feet wide, and seed lightly covered. The cultivation was the same as that given to corn, after thinning it to a stand of one stalk to every 4 inches.

The storm of the 19th of August completely prostrated the canes, and on September 12 the entire field was green with a luxurant growth of suckers.

The varieties grown at the North Louisiana Experiment Station, Calhoun, La., were:

1. Minnesota Early Amber seed, from Nebraska.
2. Early Amber seed, from Department of Agriculture.
3. Early Orange seed, from Department of Agriculture.
4. New Orange seed, from Kansas.
5. White India seed, from sugar experiment station.
6. Link's Hybrid seed, from sugar experiment station.
7. Golden Rod seed, from Kansas.

These were planted on April 18, thinned to a stand, and cultivated in its order with the corn crop. Here flat cultivation was exclusively practiced during the season, while at the other two stations high ridges were required for drainage.

These plantings were made with a view of testing, by mill and laboratory experiments, the adaptability of sorghum as a sugar crop to Louisiana. If sugar can be made profitably from sorghum anywhere in the United States it should be done in Louisiana. Chemical analyses show a larger percentage of sugar and a smaller quantity of glucose in sorghum grown in Louisiana than anywhere else in this country. At least the published analyses now at hand verify this assertion. Again, could our sugar planters be persuaded that sorghum could be made to yield a profitable quantity of sugar, say even 1,000 pounds per acre, they would soon adopt it as an adjunct to the cane crop. Once establish the fact that sugar can be profitably made from sorghum, and it will become exceedingly popular with all cane-growers, for the following reasons:

(1) By planting different varieties and at different times it can be made to ripen in Louisiana at any time from July to November—thus giving employment six months to an expensive machinery, which is now engaged only sixty days in grinding the cane crop.

(2) The cost of seed required to plant a crop of sorghum is very small, quite insignificant compared with the large amount required for cane.

(3) The ease and cheapness with which this crop can be grown.

(4) The value of the seed for forage—a by-product without cost, save the expense of carefully housing.

Again, there are vast tracts of rich alluvial lands in the middle and northern portions of the State which are too far north for cane and which will grow excellent crops of sorghum. These lands are now in cotton, but could it be demonstrated that they could grow sorghum profitably, central factories would spring up in every direction and this crop would supplant cotton in part, if not entirely.

With these possibilities in view the Director has persistently planted sorghum for three years upon the Sugar Experiment Station and attempted every year to make successfully sugar from it by the milling process. Chemical analyses have shown that our juices were rich in sucrose and low in glucose, but our sugar-house experiments have failed to extract it successfully. We have made the masse cuite full of grains, but our centrifugals failed to purge. All this was due to the starch present in the juice (extracted by pressure with the mill), which, during the subsequent process of concentration, was converted into dextrine, and this substance, our *bête noir*, prevented the elimination of the sugar. Our past experiments have demonstrated the inapplicability of the crushing mill to sorghum. They have also shown that high temperature must be avoided. Therefore new methods of extracting the juice and processes of cooking in vacuo must be resorted to before we can successfully extract sugar from sorghum.

Fort Scott, Kans., and Rio Grande, N. J., have both demonstrated that diffusion was applicable to the extraction of juice and goodly quantities of sugar had thus been obtained. After planting the above crops the State bureau of agriculture, which has immediate control of the stations, received a petition in the form of a series of resolutions from the Ascension Branch of the Sugar Planters' Association, asking that it make an appropriation for the purpose of erecting a diffusion battery for sorghum and to continue the experiments so auspiciously begun at Fort Scott and Rio Grande. The planters were anxious to know if the flattering results obtained in Kansas could not be realized here. The bureau having received at one time the deferred half of the annual Hatch appropriation, decided to grant the request so far as the limited means at their disposal would permit. Accordingly it passed a series of resolutions appropriating money for the enterprise and authorizing the Director to proceed at once to obtain the necessary machinery.

As soon as these resolutions were passed increased areas were planted in sorghum at each station, using seed received from Kansas at Kenner, and Early Amber and Orange at the other two.

Acting under these resolutions, bids were invited for building first a "diffusion battery of 14 cells, capacity of battery $1\frac{1}{2}$ to 2 tons per hour; second, a double effect of 400 square feet of heating surface. Messrs. Edwards & Haubtman, of New Orleans, making the best proposition for the erection of above machinery, were accorded the contract.

Mr. J. P. Baldwin, of St. Mary's Parish, who had formerly been an attaché of the station, and who has great mechanical ingenuity, was employed in May to superintend the erection of the machinery, and after full and free conference with him and Mr. E. W. Deming, late engineer in charge of the Fort Scott sugar works and now supervising engineer of the Conway Springs sugar works, Kansas, the following machinery was ordered: Cutter and comminiutor or pulper, with shafting and pulleys, from George J. Fritz, Saint Louis, Mo.; conveyors, elevators, and gearing from Link Belt Company, Chicago, and Mr. E. W. Deming kindly superintended the construction of a fan, a duplicate of the one made for Conway Springs sugar works, which he shipped us from Kansas.

Considerable work had to be done to conform the old sugar-house to its new machinery. Indeed the task of planning and transforming the old conditions to the new was one requiring patience, energy, and excellent mechanical ingenuity. That it has been well done is the universal testimony of all visitors.

After the above work had been contracted for, the gratifying intelligence was received from the Hon. Norman J. Colman, Commissioner of Agriculture, Washington, D. C., that he would allow this station $5,000 of the $100,000 recently appropriated by Congress for experiments in making sugar from sorghum. This supplement to the appropriation from the Bureau of Agriculture has enabled the station to enlarge its equipment and extend its field of investigation.

From our past experience with sorghum it was inferred that our crop planted on the 16th of April would not be ready for the sugar-house before 1st of September. Accordingly we contracted with Messrs. Edwards & Haubtman to deliver the machinery by the 15th of August, thus giving us fifteen days (ample time) for its erection and preparation for work. Messrs. Edwards & Haubtman failed to deliver until the 23d instant, which failure, in connection with the unprecedented storm of the 19th instant, which prostrated completely our sorghum, proved most disastrous to our successful manufacture of sugar.

In 1886, sorghum planted April 5, was harvested September 13. In 1887, sorghum planted April 21, was worked up September 23. Both years they were worked at full maturity, excepting the Early Amber and Chinese, which were ripe in July of each year.

It was fair therefore to calculate that, without any natural intervention, the sorghum this year would not be ready for the sugar-house before the middle of September; and had not the storm prevailed the date of delivery of Messrs. Edwards & Haubtman would have still afforded us ample time to have completed erection before the maturity of the crop. Either alone would not have proven disastrous; both together were fatal. [See chemical analysis further on for verification.]

Of the varieties mentioned above, the Ambers were ripe in July, and accordingly were worked up by the mill, cooked to masse cuite and left in hot room for comparison with masse cuite from diffusion juice.

LABORATORY WORK.

During the summer the laboratory has been engaged in the study of the chemistry of sorghum. To this end weekly analyses of all varieties have been made and daily study prosecuted as to the physiological changes occurring in the growth and maturity of sorghum. The following are the notes made by my assistant, Mr. W. L. Hutchinson, up to September 1, at which time he resigned to accept the professorship of chemistry in the Agricultural and Mechanical College of Mississippi. His leaving put an end to his interesting investigations.

June 21.—Iodine shows no starch in Minnesota White Amber, just headed. Single polarization gives no sucrose.

The following were found: Glucose, 3.65 per cent.; solids, 6.66 per cent.; albuminoids, .17 per cent.

The precipitate produced by subacetate of lead, after being freed from the lead, gave no trace of oxalic acid, but a quantity of tartaric acid. So great was the latter that every attempt at its entire removal failed, so that no positive conclusions as to the other acids present were drawn.

On July 16 fully matured samples of Early Amber were obtained, the juice extracted and subjected to analysis. The sucrose was determined by single and double polarization and by Fehling's solution. The following are the results:

Sucrose: Total solids, 16.58; single polarization, 12.31; double, 12.28; Fehling's, 12.22. This juice was concentrated to sirup, and the latter gave, by single polarization, sucrose, 52.41; double polarization, 53.58.

STARCH IN SORGHUM.

With green canes just heading no indications of starch are given by iodine. If there were any blue it was completely obscured by the intensely brown coloration. This brown coloration indicated dextrine and other forms of soluble starch.

With well-matured canes iodine gives an intensely blue color towards the top, decreasing in intensity towards the butt. Canes occupying an intermediate condition between these extremes, or in that stage of growth when maturity begins to appear, as indicated by the presence of sucrose in the lower part of the stalk, starch will be found in the butt but not in the top.

The above conclusions of Mr. Hutchinson have been fully confirmed by subsequent experiments; and it is not unusual in our laboratory now to prognosticate the amount of sucrose in a cane by the presence of starch, so intimately are they associated. Both sucrose and starch seem to be formed simultaneously—the former from glucose and perhaps other bodies, and the latter from dextrine and other soluble forms.

Glucose occurs in largest quantities when the polariscope gives no indication of sucrose by single polarization. In a sample of green cane, in which there was no starch and by single polarization no sucrose, but by double polarization 1.53 per cent., as high as 7 per cent. of glucose was found. As the cane from which the above sample was selected, matured, repeated analyses made at short intervals showed that the glucose decreased, until at maturity it reached as low as 0.8 per cent.

SINGLE VS. DOUBLE POLARIZATION.

In juices from matured canes there is a very close agreement between the sucrose obtained by single and double polarization. Not so with the immature canes, and the greater the immaturity the greater the disagreement. In all of the laboratory work on samples taken from the field sucrose was therefore determined by single and double polarization.

ANALYSES OF VARIETIES OF SORGHUM.

These were begun July 11 and continued weekly until worked up. The following table gives the results:

Analyses of the varieties of sorghum at different stages of growth, Sugar Experiment Station, Kenner, La.

Date of analysis.	Variety.	No. of experiments.	Total solids.	Sucrose.		Glucose.
				Single polarization.	Double polarization.	
July 11	Early Orange	19	9.8	2.2	3.22	2.95
Aug. 6	.....do	19	16.6	12.4	12.40	1.00
Aug. 13	.....do	19	16	12.3	12.60	.76
Aug. 20	.....do	19	16.5	12.1	12.21	.60
Aug. 27	.....do	19	16.3	12.2	12.52	.73
Sept. 4	.....do	19	15.7	11.7	12.85	1.23
Sept. 8	.... do	19	14.5	10.2	.....	1.05
July 11	Link's Hybrid	18	11.5	5.2	6.22	3.20
July 19	.....do	18	12.68	8.3	......	1.64
Aug. 6	.....do	18	16.20	12.2	12.10	1.28
Aug. 13	.....do	18	13.20	10	10.06	1.27
Aug. 20	.....do	18	16.10	12	12.07	.74
Aug. 27	.....do	18	16.20	12	12.18	.86
Sept. 4	.....do	18	15.30	12	12.00	.95
Sept. 12	.....do	18	11.40	7.9	......	.99
July 11	Kansas Orange	17	11.80	4.1	5.12	3.40
Aug. 6	.....do	17	16.00	12.0	12.00	1.13
Aug. 13	.....do	17	15.60	11.6	11.63	1.45
Aug. 20	.....do	17	16.80	11.7	11.67	2.78
Aug. 27	.....do	17	15.20	11.1	11.33	1.33
Sept. 4	.....do	17	13.70	9.7	9.67	1.98
Sept. 12	.....do	17	11.60	8.1	......	1.43
July 11	Early Amber, Nebraska ..	16	13.30	8.3	8.15	2.85
July 19	.....do	16	15.70	12.1	......	1.20
July 26	.....do	16	14.80	11.0	......	1.18
July 30	.....do	16	17.20	12.3	......	1.74
July 11	Early Amber, Kansas	15	13.60	8.4	9.20	2.75
July 26	.....do	15	15.70	12.0	......	1.13
July 30	.....do	15	16.73	12.1	......	1.70
July 11	Early Amber, Department of Agriculture	14	13.2	7.0	7.78	3.71
July 26	.....do	14	17.5	13.5	......	1.00
July 30	.....do	14	16.3	11.6	......	1.59
July 11	Texas Honey Drip	12	8.9	.....	1.53	6.34
July 20	.....do	12	10.57	3.3	......	4.85
Aug. 6	.....do	12	12.10	5.8	5.41	2.99
Aug. 13	.....do	12	11.9	7.9	8.25	2.20
Aug. 20	.....do	12	14.3	9.5	9.79	2.51
Aug. 27	.....do	12	13.2	9.3	9.25	2.78
Sept. 4	.....do	12	12.8	9.5	9.53	2.78
Sept. 12	.....do	12	10.4	7.7	......	2.17
July 11	Honey Drip	11	11.1	6.2	8.89	1.70
July 20	.....do	11	11.01	5.0	......	4.25
Aug. 7	.....do	11	10.1	5.8	5.41	2.69
Aug. 13	.....do	11	11.8	7.9	8.25	2.20
Aug. 20	.....do	11	11.8	6.6	6.03	1.97
Aug. 27	.....do	11	14.9	11.0	11.08	.80
Sept. 4	.....do	11	8.6	5.5	5.43	1.47
Sept. 12	.....do	11	9.3	4.9	......	2.22
July 11	Golden Rod	10	8.5	2.0	4.18	3.40
July 20	.....do	10	....	6.5	......	2.00
Aug. 7	.....do	10	13.6	8.0	8.71	1.63
Aug. 13	.....do	10	13.3	7.0	7.30	2.45
Aug. 20	.....do	10	11.7	6.3	6.50	1.21
Aug. 27	.....do	10	10.2	5.5	6.05	.81

Analyses of the varieties of sorghum at the different stages of growth, etc.—Continued.

Date of analysis.	Variety.	No. of experiments.	Total solids.	Sucrose. Single polarization.	Sucrose. Double polarization.	Glucose.
Sept. 4	Golden Rod	10	10.2	5.6	5.02	1.47
Sept. 12	...do	10	9.5	4.9	......	2.35
July 11	New Orange	9	13.3	6.9	8.81	4.25
July 20	...do	9	16.3	11.0	......	2.83
Aug. 7	...do	9	13.80	10.3	10.36	1.68
Aug. 13	...do	9	12.50	8.8	8.92	1.71
Aug. 20	...do	9	12.20	6.9	7.33	2.94
Aug. 27	...do	9	12.20	8	8.16	2.82
Sept. 4	...do	9	10.20	6.2	6.20	2.68
Sept. 12	...do	9	9.10	7.1	......	2.54
July 11	Kansas Orange	8	10.60	4.8	6.07	2.68
July 20	...do	8	13.11	8.2	......	2.21
Aug. 7	...do	8	13.90	8	8.8	1.83
Aug. 13	...do	8	14.8	10.6	10.74	1.36
Aug. 20	...do	8	12.7	8.1	8.35	1.37
Aug. 27	...do	8	13.1	7.9	8.00	1.71
Sept. 4	...do	8	10.1	6.5	6.74	2.15
Sept. 12	...do	8	......	5.3	......	1.60
July 11	Early Orange	7	11.7	6.0	7.51	2.43
July 20	...do	7	11.71	7.2	......	2.21
Aug. 7	...do	7	11.0	7.8	7.10	1.77
Aug. 13	...do	7	......	11.0	11.18	1.90
Aug. 20	...do	7	14.3	9.0	9.31	1.71
Aug. 27	...do	7	12.3	9.5	9.40	1.72
Sept. 4	...do	7	10.9	7.1	7.21	1.92
Sept. 12	...do	7	8.1	4.9	......	1.95
July 11	Enyama	6	9	2.3	3.95	2.12
July 20	...do	6	9.71	4.8	......	2.31
Aug. 7	...do	6	14.80	10.0	10.80	1.14
Aug. 13	...do	6	13.20	9.0	9.18	1.43
Aug. 20	...do	6	14.70	10.6	10.88	1.08
Aug. 27	...do	6	14.60	10.5	10.50	.82
Sept. 4	...do	6	8.5	5.2	5.05	1.47
Sept. 12	...do	6	....	6.5	......	.54
July 11	White India	5	10.0	5.4	6.9	1.82
July 20	...do	5	14.83	11.0	......	1.70
Aug. 7	...do	5	14.60	10.2	11.0	1.14
Aug. 13	...do	5	13.50	9.5	9.9	1.59
Aug. 20	...do	5	10.30	6.6	7.01	2.36
Aug. 27	...do	5	13.6	9.20	9.18	.72
Sept. 4	...do	5	13	9.90	9.80	1.27
Sept. 20	...do	5	14.1	10.00	......	1.25
July 11	White Mammoth	4	6.5	.4	2.09	3.29
July 20	...do	4	7.91	2.6	......	3.00
Aug. 7	...do	4	14.20	9.6	9.71	1.43
Aug. 13	...do	4	10.5	6.0	6.40	2.30
Aug. 20	...do	4	10.2	6.1	6.54	1.87
Aug. 27	...do	4	12.2	7.7	7.84	.87
Sept. 4	...do	4	8.1	5.7	5.06	2.00
Sept. 20	...do	4	10.5	6.9	......	2.14
July 11	Link's Hybrid	3	9.8	4.8	5.78	1.59
July 20	...do	3	9.1	4.0	......	2.55
Aug. 7	...do	3	14.9	9.0	9.53	2.34
Aug. 13	...do	3	14.5	10.1	10.21	.74
Aug. 20	...do	3	13.7	9.2	9.55	1.14
Aug. 27	...do	3	13.7	10.5	10.50	.78
Sept. 4	...do	3	12.2	9.1	9.10	1.00
Sept. 20	...do	3	10.6	6.7	......	1.48
July 11	Honduras	2	7.0	2.0	2.06	1.0
July 20	...do	2	7.81	3.4	......	3.00
Aug. 7	...do	2	9.70	3.6	4.80	2.14
Aug. 13	...do	2	7.10	3.4	3.52	2.76
Aug. 20	...do	2	7.70	2.5	3.05	2.53
Aug. 27	...do	2	......	7.1	7.12	1.94
Sept. 4	...do	2	7.6	5.0	4.99	2.11
July 11	...do	1	6.8	1.0	1.81	3.40
July 20	...do	1	8.81	4.4	......	3.09
Aug. 7	...do	1	10.60	6.2	7.79	1.83
Aug. 13	...do	1	9.20	5.8	5.83	1.50
Aug. 20	...do	1	9.20	4.0	3.87	3.14
Aug. 27	...do	1	10.50	0.6	6.82	1.79
Sept. 4	...do	1	8.0	5.4	5.46	1.74
Sept. 12	...do	1	10	6.0	......	2.27

Analyses of varieties of sorghum grown at Baton Rouge, La.

Date of analysis.	Variety.	Total solids.	Sucrose, single polariza-tion.	Glucose.
Aug. 6	Early Amber		12.00	
Aug. 9	......do	15.9	9.50	3.80
Aug. 14	......do	18.1	13.40	1.12
Aug. 28	......do	17.0	12.10	1.09
Sept. 11	..... do	14.7	7.30	1.82
Aug. 6	Early Orange		11.20	
Aug. 14	......do	15.9	10.00	2.38
Aug. 28	......do	17.0	12.40	2.07
Sept. 11	..... do	11.9	7.8	4.52
Aug. 6	Link's Hybrid		9.4	
Aug. 9	......do	16.1	11.5	1.87
Aug. 14	......do	16.4	10.5	3.00
Aug. 6	Honduras		6.3	
Aug. 9	......do	15.8	8.4	4.70
Aug. 14	......do	11.6	4.1	5.47

Analyses of varieties grown at North Louisiana Experiment Station, Calhoun, La.

Date of analysis.	Variety.	Sucrose, single polariza-tion.	Glucose.
Oct. 1	Early Amber	11.4	1.27
Oct. 1	Early Orange	11.8	2.56
Oct. 1	New Orange	10.5	2.20
Oct. 1	Link's Hybrid	12.3	1.56
Oct. 1	White India		.87
Oct. 1	Golden Rod	10.6	1.36

An inspection of above tables will show that Early Amber reached its maximum in July, say one hundred days after planting. Golden Rod and Honduras never reached maturity, the storm of the 18th prostrating them before the maximum of sugar was reached. The other varieties attained their maximum during August.

Could these experiments have been worked during August, it is believed that most excellent results would have been attained. Up to September 4, just as suckers began to appear at each joint on the prostrate cane, the latter had lost but little in sucrose since the storm of the 19th. After the suckers began to grow the loss was rapid and heavy, as is shown by the mill juices of September 8 to 20.

The canes at Calhoun were not injured, the storm not extending as far north as this station. They have therefore preserved their sugar up to October 1 and suffered little or no loss.

EXPERIMENTS IN DIFFUSION.

All the machinery being in position and ready for use, a trial run was made on September 8, using the Early Orange variety. The cutters did their work well; so did the diffusion cells, except now and then a leak which was easily closed. The larger heater, which heated the juice before entering the cells, was out of order and could not be used either in this or the next trial. The fan which had been furnished as adapted to the cleaning of sorghum chips failed utterly to do its work. The shaker which was geared to the fan ran too rapidly, and had to be run by an independent pulley at a slower motion. The depth of the shaker was far too narrow, so much so that the chips of cane thrown violently forward by the force of the cut were often propelled beyond the shaker and fell into the trash. In this way a large amount of the cane in this experiment was lost. The shaker was lengthened and many other improvements made until good work was accomplished. On account of these defects only 1,152 pounds of sorghum, with tops and blades, were used and only two cells of the battery were filled. The following are the laboratory analyses:

	Total solids.	Sucrose.	Glucose.	Ratio of sucrose to glucose.
Mill juice.....................	14.6	10.2	1.05	10.25
Diffusion juice:				
First cell		1.1	.1021	9.11
Second cell		.7	.0638	9.11

No sugar or sirup made.

Pending the making of the necessary improvements to the fan and shaker the cubical contents of the cells were carefully calculated in the following manner: The cells were filled with water and then the water carefully emptied into a sugar wagon and weighed, allowing 62½ pounds of water to a cubic foot. Each cell contained 13.52 cubic feet. A cell packed with sorghum chips and one put in without packing were also emptied and weighed. Their weights were, respectively, 353 pounds and 276 pounds, making 26 pounds and 20 pounds per cubic foot.

Without entering into the full details of daily work, the following, taken from our large amount of records, will suffice to illustrate fully the work performed.

Considering the very low character of the sorghum worked, the results obtained are quite promising.

Monday, September 10, 1888.—Another trial of the machinery was made to-day to decide whether the improvements so hastily made were effective. Honduras sorghum was used; weight, with tops and blades, 2,158 pounds. Everything worked fairly well. It was found that both the cutter and comminutor were projecting the chips in every direction, thus causing great waste. A stop was made and these boxed in. Four cells were, however, filled, and the juices from these concentrated in the double effect and left in the latter all night. The next morning, to our surprise, we found that one of the tubes of the double effect had leaked during the night and had diluted the sirup almost to the original juice. Accordingly it was withdrawn and thrown away, and the leaking tube plugged up. The laboratory results are given:

	Sucrose.	Glucose.	Ratio sucrose to glucose.
Milk juice..................	4.3	2.51	58.3
Diffusion juice—			
First cell..............	1.3	.43	....
Second cell..........	1.3	.38	....
Third cell	2.3	.76	....
Fourth cell·..........	1.4	.55	....

Wednesday, September 12.—Having repaired the defects, work was begun at 9.30 o'clock and continued until nineteen cells had been filled. Everything worked admirably except the heaters, which were not under control, and hence varying temperatures used in diffusing. Weather very warm and much suffering experienced by everybody at work, particularly by the men at the diffusers and clarifier.

The following canes, with quantities, were used :   .

	Pounds.
Link's Hybrid, with tops and blades..........	1,292
Kansas Orange, with tops and blades.........	900
Texas Honey Drip, with tops and blades.......	1,214
Honduras...................................	470
Honey Drip.................................	828
Golden Rod.................................	1,096
New Orange................................	1,072
Kansas Orange	829
Early Orange	1,370
Total	9,071
Less tops, 1,403 pounds } =28.46 per cent... Less trash, 1,170 pounds	2,582
Clean cane diffused	6,489

The chips packed in very tightly failed to discharge easily. Drew the first juice off at cell No. 7, and continued to draw until twenty-five discharges had been made, viz, Nos. 7, 8, 9, 10, 11, 12, 13, 14, 1, 2, 3, 4, 5, 6, 8, 9, 10, 11, 12, 13, 14, 1, 2, 3, 4, 5.

The juice from No. 7 passed over seven fresh chips.

The juice from No. 8 passed seven 2d chips and one fresh chips.

The juice from No. 9 passed over seven 3d chips, one 2d chips, and one fresh chips.

The juice from No. 10 passed over seven 4th chips, one 3d chips, one 2d chips, and one fresh chips, etc., until the 14th cell was reached. While No. 14 was being filled No. 1 was emptied. Then began regular diffusion. The 20th cell was partially filled but not used, and No. 21 was at the same time emptied. Hence the absence of Nos. 6 and 7 in the discharges above.

The following analyses were made :

```
1. Mill juices of each variety used.
2. Diffusion juices from each cell.
3. Chips as they were emptied from each cell.
4. Clarified juice from each clarifier.
5. Sirup.
6. Residuum scums.
7. Sugar.
8. Molasses.
```

The following are the results :

Mill juices.

Variety.	Total solids.	Sucrose.	Glucose.
Link's Hybrid	11.4	7.9	.99
Kansas Orange..............	11.6	8.1	1.43
Honey Drip..................	10.4	7.7	2.17
Honduras	10.0	6.0	2.27
Golden Rod.................	9.5	4.9	2.38
New Orange	9.1	7.1	2.54
Kansas Orange..............		5.3	1.60
Early Orange	8.1	4.9	1.95
Enyama.....................		6.5	.54

Diffusion chips.

1	1.4	10 with eight washings...	.8
2	.6	11 with seven washings...	.7
3	.5	12 with six washings	.5
4	.6	13 with five washings.....	.6
5	.2	14 with four washings	1.2
6 with twelve washings.....	.55	15 with three washings....	.7
7 with eleven washings.....	.75	16 with two washings. ..	1.5
8 with ten washings	.85	17 with one washing.	(*)
9 with nine washings.......	1.10		

* Sample lost.

Diffusion juices.

	Total solids.	Sucrose.	Glucose.	Glucose ratio.
First discharge.....................	6.4	4.2	1.11	26.45
Second discharge..................	5.5	3.8	1.12	26.45
Third discharge............,......... ⎫ Fourth discharge................ ⎭	4.1	3.0	.53	17.67
Fifth discharge................. ⎫ Sixth discharge................ ⎭	4.1	3.1	1.19	38.39
Seventh discharge................	5.9	3.8	1.56	41.05
Eighth discharge.................	5.1	3.7	1.40	37.84
Ninth discharge..................	5.6	3.9	1.39	35.64
Tenth and subsequent discharges	4.7	3.3	1.56	47.27

Clarified juices.

1	4.5	3.4	1.06	31.18
2	4.9	3.3	1.26	38.18
3	2.8	2.2		
4	2.2	1.7	.65	38.23

Sirup:		Scums:	
Total solids........	32.20	Total solids	4.10
Sucrose..................	17.50	Sucrose.................	1.90
Glucose..................	7.35	Glucose............. ..	.83
Glucose ratio	42.00	Glucose ratio	43.68

Sugar :		Molasses :	
Sucrose	91.2	Sucrose	30.4
Glucose......	2.85	Glucose	14.28

It was utterly impossible, from the varying amounts of sucrose in the canes used, to get anything like uniform results either on the juices or chips. There were drawn four clarifiers, of about 500 gallons each. The last two were very dilute, owing to the excess of water used in washing the chips after cells were filled. This juice was heated with lime and brought to neutrality ; heated, and blanket, which was quite insignificant, removed. It was then settled and clear juice run into the double effect and concentrated.

There was a large quantity of settlings and some scums, which were weighed and analyzed and thrown away to avoid interfering with the well-clarified sirup. The following are weights obtained :

	Pounds.		Pounds.
Sirup.................	1,562	Sugar	49
Settlings and scums	1,070	Molasses	752

The following are the notes of diffusion :

Every effort was made to hold the temperature at 200° Fahrenheit, but until the battery had been used in one entire round this is almost impossible to do, since sending in quickly water heated to 200° Fahrenheit into cold iron cells filled with cold chips the loss of heat by radiation and convection is very great. Six minutes were allowed for the diffusion of each cell after the hot water was turned on. Every effort to grain in the vacuum pan proved abortive, as the following notes of Mr. Baldwin, who had charge of the pan and was assisted by Mr. Barthelemy, will show :

"Part of juice concentrated in double effect on first watch, remainder on second watch, when the juice got very hot, 180°, and was emptied in cars to cool; finished concentrating on morning of 13th at a temperature of 155° to 160° Fahrenheit. Juice dark colored and some feculent matter present. After mixing sirups started vacuum strike pan at 2 p. m. on 13th; temperature, 138° to 140° Fahrenheit ; very thick ; nothing but candy would form in the pan. Allowed to stand half an hour until candy dissolved, but no grain. Stood again one hour; at 7 p. m. still no grain. Cooked very thick and remained in pan until 2 p. m. next day, when it was all boiled to string sugar and put in the hot room. Injured some by being cooked to candy.

"In the hot room it began at once to grain, until the wagon was quite solid with small grains of sugar.

It was centrifugaled and gave the following results :

	Pounds.
Sugar	49
Molasses	752

RECAPITULATION.

	Sucrose.		Sucrose.
	Pounds		*Pounds.*
Cane contained	349.75	Chips contained	56 20
Syrup contained	273.22	Sugar contained	34.58
Scums contained	20.33	Molasses contained	228.61

	Pounds.
Sugar obtained per ton sorghum	12.5
Molasses obtained per ton sorghum	237.1

After the analyses of the mill juices were known, little or no hope was entertained of successful sugar results. Indeed, it is wonderful with such juices and after such treatment that any sugar should be obtained.

September 17.—It has often been published that neither sorghum nor its juices will stand transportation or delay in working them up, after being cut. That such is not the case with us is abundantly proved by the following and many other experiments during this season : On September 16 Mr. Barrow, assistant at the State experiment station, was sent to Baton Rouge to harvest and ship a car-load of sorghum from that station to this. By 9 a. m. on the morning of the 16th he had cut and loaded a closed car with Early Orange sorghum. This sorghum was quite wet from dew and had its leaves and tops still on—conditions making fermentation quite feasible to almost any crop. It was delivered at Kenner by Mississippi Valley Railroad at 7 p. m. of same day. It was unloaded and delivered at sugar-house at 12 m. of the 17th, and worked up as delivered. The cane had been badly blown down by the storm of the 19th, and was filled with suckers several feet long, now in full heads. It was quite low in sugar, as the following analysis of selected stalks, made on September 11, showed :

Total solids	11.9
Sucrose	7.8
Glucose	4.52

Began diffusion at 9 a. m. Filled twenty-three cells with chips and drew off thirty-one cells of juice. Finished in early evening, after two slight detentions. Cells diffused sixteen minutes each, except three times, when interrupted. The temperature varied from 150° to 200° F. The juice was boiled to a sirup in double effect and made into string sugar in the vacuum pan. Boiled all night, finishing the next day. The string sugar was run into the hot-room, where it was grained into almost a solid mass. The following are the amounts used:

```
Weight of canes ............................................................................. 13,266
Less weight of tops ..........................................................  2,445
Less weight of leaves .......................................................  1,765
Less weight of trash in yard ...............................................  1,558
Less weight of chips not used .............................................     82
                                                                         ——— 5,867
```

```
    Clean cane used .......................................................... 7,399
```

The juices from this were concentrated into a sirup, giving 1,491 pounds; scums thrown away, 313 pounds; juice made into molasses, 259 pounds.

The following are the laboratory results:

```
                                                                          Pounds.
Sugar obtained ................................................................ 115
Molasses obtained ............................................................ 672
Sugar per ton of sorghum .................................................. 31.4
Molasses per ton of sorghum ............................................. 181.8
```

RECAPITULATION.

```
Cane contained (calculated) ..............................pounds sucrose.. 435
Sirup made into sugar contained .........................do........ 328
Sirup made into molasses contained ....................do........  57
Scums contained .............................................do........   7
Chips contained ..............................................do........  32
Fiber in cane ................................................per cent.. 15.5
```

Early orange sorghum.

	Total solids.	Sucrose.	Glucose.	Glucose to sucrose.
				Per cent.
Mili juices .... {	11.4	7.0	3.33	48
	11.3	7.0	3.58	51
	11.7	6.9	3.30	48
	....	3.2	1.79	.56
	....	3.95	2.00	.51
	....	3.00	1.92	.64
	....	3.90	2.17	.55
	....	3 90	2.32	.59
	....	4.10	2.00	.58
Diffusion juices .... {	....	3.50	1.72	.49
	....	3.70	1.46	.39
	....	4.10	1.73	.42
	....	3.50	1.50	.48
	....	3.60	1.66	.46
	....	4.20	1.62	.38
	....	3.90	1.70	.44
	....	3.30	1.60	.48
	....	.3	.14	.47
	....	.3	.18	.60
	....	.25	.16	.64
Diffusion chips .... {	....	.35	.149	.43
	....	.25	.14	.56
	....	.15	.13	.90
	....	.15	.10	.40
	....	3.6	1.85	51
	....	3.9	1.60	41
Clarified juices .... {	....	3.1	1.57	51
	....	1.8	.99	55
	....	1.3	.56	43
	....	1.1	.54	49
Sirup	....	22	11.1	50
Scums	....	4.2	2.22	53
Sugar	....	92.1	2.94	....
Molasses	....	34	22.72	....

Here, as before, the dilution was great, owing to the water used in washing the chips after cells were filled. This cane had nearly a constant composition, and from glucose ratio there has been little or no inversion either in cells or in concentration of sirup. In fact, when water at 200° F. is sent into cells and maintained there for six minutes at this temperature little or no inversion took place, notwithstanding the weather gauge showed this day a maximum temperature of 83° F.

September 20.—The following canes were selected for this run : Link's hybrid, White India, White mammoth, and the second planting of Early Amber. The suckers, of which there were many, were removed by hand. Filled nine cells. Everything worked well.

Pounds.

Weight of cane used	5, 078
Less weight of tops	812
Less weight of trash	653
Less weight of suckers	208
Less weight of chips not used	74
	— 1, 747
Clean cane used	3, 331

Juice neutralized with lime, blanket removed, settled, concentrated in double-effect and *grained* in vacuum pan; then emptied into car and run into hot-room, where it solidified into crystals of sugar of small size.

Pounds.		*Pounds.*	
Weight of sirup	695	Weight of molasses	235
Weight of scums, etc	150	Sugar, per ton	24
Weight of sugar	40	Molasses, per ton	141

The following are laboratory results :

Variety.	Mill juices.				Diffusion juices.				Diffusion chips.	
	Total solids.	Sucrose.	Glucose.	Glucose to sucrose.	Total solids.	Sucrose.	Glucose.	Glucose to sucrose.	Sucrose.	Glucose.
				Pr. ct.				*Pr. ct.*		
Link's hybrid	10. 6	6. 7	1. 48	22	4. 8	3. 05	1. 13	.37	.20	.16
White India	14. 1	10. 0	1. 25	12½	6. 0	3. 50	1. 51	.43	.30	.14
White mammoth	10. 5	6. 9	2. 14	33	6. 0	3. 70	1. 51	.41	.20	.13
White amber (Nebraska) ..	10. 7	6. 5	1. 92	29	5. 2	3. 20	1. 57	.40	.10	.12
White amber	10. 4	5. 4	3. 12	57	5. 6	3. 25	1. 61	.49	.10	.12

Clarified juice :			Sirups—Continued :	
Total solids	{ 5. 9		Glucose to sucrose..per cent	40
	{ 2. 1		Scums :	
Sucrose	{ 3. 5		Sucrose	1. 7
	{ 1. 4		Glucose	.73
Glucose	{ 1. 39		Glucose to sucrose..per cent	41
	{ .51		Sugar :	
Glucose to sucrose..per cent	{ 39		Sucrose	92. 2
	{ 38		Glucose	2. 93
Sirups :			Molasses :	
Total solids	32. 94		Sucrose	34
Sucrose	17. 5		Glucose	20
Glucose	7. 04			

RECAPITULATION.

Sucrose in sirup	121. 62	Sucrose in sugar made	36. 88
Sucrose in scums	2. 55	Sucrose in molasses made	79. 90
Sucrose in chips	16. 56	Fiber in cane..per cent	15. 04

The following determinations of albuminoids were made : .

Mill juices :		Diffusion juices :	
Link's hybrid	.430	Sept. 12	.0531
Kansas orange	.215	Sept. 17, Baton Rouge cane	.0748
New orange	.322	Sept. 20	.1276
Early orange	.425	Clarified juices :	
Early orange, Baton Rouge	.371	Sept. 12, first clarifier	.0319
Do	.345	Sept. 12, second clarifier	.0212
Mill juices for Sept. 20	.307	Sept. 17, Baton Rouge cane	.0357
		Sept. 20	.0843

It will thus be seen that diffusion juices contain much less albuminoids than mill juices.

LATE PLANTING OF SORGHUM.

After deciding to erect a diffusion battery to work up sorghum, a
late planting was made upon land from which a crop of oats had been
harvested. The oats were harvested May 15, and the land broken with
four-horse plow and harrow. Sorghum planted May 23. The continued
rains during June and July prevented necessary cultivation. The storm
of August 19 prostrated it, and, though far from being ripe, never re-
covered. Most of these seed were received from Mr. William P. Clem-
ents, of Sterling Sugar Works in Kansas, and were mainly hybrids of
different varieties. They were carefully followed during maturity with
analyses, but at no time did any of them show a large sugar content.
The following table will show analytical results:

Analyses of sorghum planted May 23.

No.	Variety.	Analyzed October 8.			Analyzed September 15.		
		Brix.	Sucrose.	Glucose.	Brix.	Sucrose.	Glucose.
1	Honduras, grown in Louisiana	6.2	1.1	2.52	7.4	.1	4.13
2	White Amber, grown in Nebraska	......	....	....	12.0	7.7	2.09
3	Early Amber, grown in Kansas....	......	....	....	9.3	8.4	2.28
4	Early Amber, from Department of Agriculture......................	......	....	....	7.6	6.7	2.36
5	Golden Rod, from Sterling, Kans..	9.2	?.7	2.73	9.3	3.7	2.12
6	New Orange, from Sterling, Kans..	11.0	4.7	3.57	12.3	4.3	3.45
7	White India, from Sterling, Kans..	10.8	5.3	2.34	11.3	7.0	1.46
8	Early Orange, from Department of Agriculture...................	10.3	5.5	2.28	8.4	1.2	2.46
9	Chinese Sugar Cane, from Depart- ment of Agriculture...........	9.5	3.4	2.32	......	....	....
10	Early Orange, from Department of Agriculture.....................	7.0	2.3	2.13	7.4	2.2	3.12
11	Hybrid, Sterling, Kaus	8.8	3.2	1.93	9.0	2.9	2.30
12	Do	11.8	6.3	2.45	7.2	4.1	3.03
13	Do	12.0	4.4	3.12	11.2	5.9	2.46
14	Do	9.1	1.6	2.34	5.8	1.2	2.20
15	Do	8.0	2.6	2.80	10.8	6.2	2.70
16	Do	8.4	3.0	3.28	9.9	4.1	2.59
17	Do	4.3	.3	1.44	7.1	1.1	1.66
18	Do	8.6	3.1	2.79	6.6	2.2	1.76
19	Do	12.2	7.0	1.44	9.9	6.1	1.60
20	Do	9.1	3.9	2.79	9.3	6.4	2.74
21	Do	9.3	4.7	2.56	8.1	3.8	2.42
22	Do	10.4	5.2	2.50	7.2	2.9	2.60
23	Do	5.5	1.0	3.47	6.0	2.0	1.70
24	Do	8.9	3.4	2.17	9.3	4.5	2.20
25	Do	8.6	4.1	2.75	9.0	3.1	3.57
26	Do	8.1	3.9	2.64	6.9	1.1	.91
27	Do	6.5	2.6	1.74	6.9	1.0	1.48
28	Do	10.9	4.7	1.67	8.9	3.0	2.64
29	Do	8.2	2.5	3.55	6.1	.8	2.27
30	Do	10.6	5.3	2.59	9.2	4.2	3.07
31	Early Goose Neck, Sterling, Kans.	10.9	5.5	3.75	12.1	7.2	3.57
32	Honduras, from Arizona	10.6	5.2	4.95	10.9	4.5	2.94
33	Pierces Cross, from Sterling, Kans.	......	....	....	7.9	3.0	4.98
34	Duchess Hybrid, Sterling, Kans...	......	....	....	9.4	4.0	3.18
35	New Sugar Cane, Sterling, Kans...	......	....	....	6.4	1.6	3.33
36	Liberian, South Arizona.........	10.1	.... 3.1	4.42	9.7	3.0	4.42
37	Liberian, Missouri...............	......	....	....	8.3	1.7	4.07
38	Liberian, Texas.................	......	....	....	7.9	1.9	4.34
39	Liberian, Alabama	......	....	....	6.3	.9	3.03

The following are the descriptions of the hybrid varieties gathered
October 8:

No. 11. Panicle, black exterior, dull red interior; two distinct heads, the one full,
with black seed red tipped, the other few seeded, slightly closed heads, probably a
cross between White Mammoth or India with a black-seeded variety.

No. 12. Large heads, black and yellowish-white, fine stalks, green plumes with pinkish-white seeds. A cross, probably, between White India and an unknown variety.

No. 13. Heads large, one sleek black, the other white red. Probably New Orange and a black-seeded variety; stalks medium.

No. 14. Only one kind; black, with red openings, full seeded.

No. 15. One black, with reddish seeds; the other black, with dull white seed; Honduras and unknown variety.

No. 16. One black, with slightly reddish seed; the other, large white heads; both full seeds; stalks small.

No. 17. Black, with red seed in one, full headed; the other white seeded, few and loose.

No. 19. Three varieties, black, white, and variegated; heads few seeded; stalks small.

No. 20. Black, with bright-red seed, few; stalks small.

No. 21. Black, with yellow opening; two varieties, one black glumes with pinkish seeds, full headed; other ashy glumes, closed heads, with few seed.

No. 22. Black, with white seed, bent neck; the other dark, with pinkish seed.

No. 23. Red, with yellow openings, one dark, with pinkish seed; the other dark, with white seeds, pinkish blush.

No. 24. Dark, with pale-yellow openings. One full headed, black glumes with pinkish seeds; the other dark glumes, closed seed in an indifferent head.

No. 25. Dark; large white opening. One, black glumes with white seed; the other, black glume with pinkish seed.

Nos. 26, 27, 28, and 29 are crosses of Honduras on white varieties, with large preponderance of Honduras.

No. 30. Black, with red openings. One black, with red seed; the other black, with white seed.

October 9 a part of the above was cut and diffused, but results in sugar were *nil;* 8,482 pounds of sorghum were successfully diffused, leaving on an average less than .15 per cent. sucrose in chips, but the juice was very dilute and contained a greater quantity of glucose than sucrose. After concentration to masse cuite it was left in the hot room for several weeks, with no indication of grain.

On November 15 the late planting of Honduras, Chinese, and Golden Rod were gathered and diffused. The yields per acre were as high for the first two as 20 tons per acre; but the sugar content was very low. The following are the analyses :

	Brix.	Sucrose.	Glucose.
Mill juices :			
Honduras	5.7	.80	1.17
Chinese	8.1	2.10	2.23
Golden Rod	8.1	1.60	2.59
Mixed diffusion juice	3.4	.60	1.25
Sirup from above		4.8	5.31

Here the process of clarifying in the cell by the use of lime was tried for the first time on sorghum. A much larger quantity of lime was used than was required for cane. Results indicated that with an abundance of lime, plenty of heat, and a very fine chip a good clarification could be obtained in the cell. Further trials, however, of this process on sorghum are needed to decide fully upon its efficacy.

Since glucose was so largely in excess of sucrose no attempt was made to obtain sugar. The sirup was concentrated into molasses and sent to the molasses-tank.

CONCLUSIONS.

While the present season was in Louisiana a most disastrous one for making sugar from sorghum, yet the successful application of diffusion in the extraction of the juice from both sorghum and sugar-cane has been abundantly proven.

From sorghums of fair quaility, such as were raised on this station in 1886 and 1887, it is certain that a large quantity of sugar could be obtained. From Early Orange this year with only 7 per cent. sucrose and 3.33 per cent. glucose (glucose ratio nearly 50), 31.4 pounds sugar were obtained to ton of sorghum. This same variety showed in 1886 a sugar content of 13 per cent., with a low glucose ratio, and in 1887, a less favorable year, sugar content of 10.5 per cent. and only 13 as the glucose ratio. Could such cane have been diffused this year, a yield of fully 100 to 125 pounds per ton might with reason have been expected.

However, the station will repeat again the experiments next year, with more promise of success.

EXPERIMENTS AT CONWAY SPRINGS, KANSAS.

REPORT OF E. W. DEMING.

I have the honor to present my report as superintendent of the experiments conducted at this place the past season by your Department in the manufacture of sugar from sorghum.

The experiments were conducted in connection with the work of the Conway Springs Sugar Company.

This company was incorporated April 10, 1888, under the laws of the State of Kansas, with an authorized capital of $100,000. Its officers are G. W. Fahs, president; E. E. Baird, vice-president; G. B. Armstrong, treasurer; E. W. Deming, secretary and manager. The buildings of this company are constructed of wood. Main building 56 by 78; foot plates with cupolas for strike pan, diffusors, double-effects, and shredding room; boiler and engine house 65 by 70 feet; cutting and cleaning house, 14 by 14; tool-house, 10 by 18; oil-house, 8 by 16; office and laboratory, 16 by 30; cane-shed, 10 by 150, two floors; scale-house, 8 by 10; cooper shop, 15 by 15.

The factory was equipped with two tubular boilers of 150 horse-power each; two 30-horse-power high-speed engines; three hanging Hepworth centrifugals with mixer; one 7-foot vacuum (dry) pan from R. Deeley & Co., New York. Hot-room, with fifty sugar-wagons; Lillie double-effect from George M. Newhall & Bro., Philadelphia; diffusion battery from Shickle, Harrison & Howard Iron Company, Saint Louis; three cutters, with necessary clarifiers, skimming-pans, and storage tanks. One dynamo of 100-lamp capacity (incandescent) provided lights for the building.

Two sets rolls and a fire drier for crushing and drying exhausted chips, and one small open evaporator.

The diffusion battery consists of sixteen cells each 8 feet long and 35 inches in diameter, wrought-iron shell with similar castings, doors and counter-weights at each end, provided with solid rubber gaskets that gave satisfaction under a 30-pound per inch pressure. One heater for each cell, made of 6 inch wrought pipe containing 11 1-inch brass tubes 5 feet long; the connecting and circulating pipes were of 2½-inch wrought iron. The battery was placed in two lines of seven cells each with one across each end, and supported on wooden posts, beams, and cross-beams 8 feet from the ground; each cell would hold 1,400 pounds of chips. The cost of this battery with pipe and fittings was $5,500; its work was in every way satisfactory. The exhausted chips were discharged into a chute of sloping sides, directing them into a drag of peculiar construction, delivering them into an elevated chute from whence a cart removed them. This apparatus worked well.

The double-effects are each 4 feet in diameter and 18 feet long placed on end; each has seventy 3-inch brass tubes 8 feet long placed vertically; ends of tubes properly secured in plates, steam being admitted to the chamber about the tubes. Pumps draw the liquor from bottoms of pans, discharging at the top, passing through perforated screens to the upper plate from which it overflows a thin film of juice down the inside of all tubes alike; the evaporation occurs in the tubes; a vacuum is maintained throughout the tubes and circulating pipes. The vapor was removed at lower end of tubes, with suitable circulating pumps and a slight change in the tops to facilitate cleaning; they will not only have large capacity but unusual merit for handling sorghum juices. These pans by reason of mechanical defects not difficult to overcome and the rapid formation of scale upon the heating surface, extremely difficult to remove, caused some considerable delay to the work.

The first or second cutter, Hughes's style, consisting of two heavy balance wheels 36 inches in diameter placed 32 inches apart on a 3-inch shaft; two knives placed horizontally connected the face of the balance wheels. The dead-knife was placed 8 inches below center of the shaft, thereby making a bevel cut on the cane; space between end of drag and dead-knife 23 inches; this permitted the seed to readily escape the knives by falling into a drag. Power was transmitted by a belt, the cutters making 200 revolutions per minute, cutting into 1-inch sections a bed of cane 30 inches wide and 6 inches deep. This cutter proved deficient in both strength and capacity; one-third of the delays and losses attending the work are traced to this source. Below the cutter was a single fan 20 inches in diameter and 30 inches long, having a motion of 600 revolutions per minute. Its work was especially fine.

The two shredders were each 20 inches long and 8 inches in diameter, provided with four knives held in place by a peculiar arrangement at the ends, leaving the face of cylinder free of openings. Motion, 1,200 revolutions per minute. Doing satisfactory work.

Three clarifiers of No. 10 iron, round, 5 feet in diameter and 30 inches deep with cone-shaped bottoms; 2-inch copper coils were used. They lacked scum pockets; otherwise their work was satisfactory.

The cane shed consisted of two floors, each 10 feet wide and 150 feet long, separated one above the other by a space of 4 feet. As a means of storing cane this apparatus worked well.

An open pan, iron, of two channels each 12 inches wide and 12 inches deep and 20 feet long, filled with three-quarters inch copper coil was at first used with steam as a skimming pan to aid clarification. Later steam was dispensed with and the pan operated as a continuous flow settling tank, giving better satisfaction and suggesting a possible manner of constructing a rapid system of continuous flow settling tanks. To prepare exhausted chips for use as fuel were provided two sets of heavy iron rollers, each set composed of two rollers 12 inches in diameter and 37 inches long, placed one above the other, the upper one having a covering of flexible rubber 1 inch thick.

One fire drier consisting of a sheet-iron cylinder 12 feet long and 4 feet in diameter open at both ends. Three sets of arms connected the shell to a 3-inch shaft passing through the center. The shaft was supported by suitable boxes and cross-pieces beyond the end of the cylinder. The whole placed in brick-work, with one end 1 foot higher than the other, and heated with direct fire underneath the lower end. Six narrow shelves upon inside of cylinder served to elevate the chips to fall through an air space as the cylinder was slowly turned by means of a link belt. This carried the chips from the upper to lower end of cylinder where they were discharged.

This apparatus was operated parts of two days. The two sets of double rolls were placed about 3 feet apart; wire netting 36 inches wide of No. 20 steel wire, 8 mesh ends, lapped and wired together, passed between the rolls of both sets, returning underneath and passing around a wooden roll underneath the discharge of the drag returning the chips from underneath the battery. This netting solved the question of feeding these chips to rolls, and I believe would work equally well upon iron rolls; the water readily escapes through the netting.

The high speed at which this wire carrier and rolls must necessarily be operated, the uneven feed from chip drag, the difficulty in distributing the chips evenly upon the netting, the failure to remove more than about 40 per cent. of the water, and the inability of the drier through which the chips afterwards passed to more than warm them were considered sufficient reason for their speedy removal. There is a possible hope for better success with these rolls if the chips are taken into a large chute from which a constant, even, well-distributed feed may be furnished them; even then artificial heat would be required to remove an additional amount of moisture before good combustion is obtained. The pieces of rind or shell cross each other, forming small spaces to be filled with pith and moisture, and the spongy nature of this pith makes it tenacious of water during the process of rolling.

The chip elevator gave some trouble when permitted to get out of repair. The wagon, turn-table, cane shed, outside drags, engines, pumps, dynamo, and strike pan gave entire satisfaction.

The centrifugals did excellent work even upon the worst melados. The process of work is as follows:

The cane is received from the farmer upon specially constructed racks. The wagon is driven on a turn-table by which it was squared about, then backed a few feet against an ordinary wagon scales on which was a raised platform 3 feet high; an iron hook was secured in the two ropes placed around the load by the farmer; a friction clutch at the opposite end of the cane shed, nearly 200 feet distant, drew the load over the rear end to the scales. Here it was weighed net, and the farmer's ropes removed. An endless sling was then thrown over the cane, the same power taking it into one of the floors comprising the cane shed, where it was left for night run or taken directly through to a small downward incline, where two men pulled it apart, feeding to three chains with attachments that carried it 1 foot above a cross drag leading to the cutters. The feed was regulated by stopping and starting this chain. This drag leading to first cutter has a motion of 40 feet per minute, carrying the cane in bundles a few inches of space between the tops of one bundle and the tops of the next; this permitted seed to drop freely. Seed was hauled directly to the field and left in small piles; that required for sugar work next season is carefully selected by hand, tied up into bundles of 18 tufts, two bundles then tied together and so hung up in a dry place. The rest is stacked, allowed to pass through a sweat, and thrashed in February. It is sold in large quantities at good prices to ranchmen, who sow it for fodder for stock. The inch sections of cane as they are cut fall into a strong blast of air directly underneath, by which the leaves and sheaths are removed. By means of a link-belt drag the cleaned sections are conveyed into the main building to an elevator, taking them above the roof, where they are discharged into the hopper of the shredder and reduced to pulp, which falls into a carrier passing over the diffusion battery. Openings in bottom of this carrier permit the cane chips to be spouted to cells on either side.

About September 15 a trial run was made with whiting (carbonate of lime) by placing it in each cell of chips, its object being to prevent inversion during the process of diffusion. The results were disappointing. At the instigation of Dr. H. W. Wiley an apparatus was provided for dusting finely powdered air-slaked lime upon the chips as they left the shredder, about 1 quart being required for each 1,400 pounds of chips. As this apparatus was under nearly perfect control, any degree of acidity of the juice desired was secured; it was generally carried nearly to the neutral point, preventing all inversion, which the whiting failed to do.

Ordinary clarifiers of 450 gallons capacity were used and the acid in juice nearly if not quite neutralized. If the juice was properly limed in the cell, very few scums were found in the clarifiers. The battery was operated at a temperature of 180° Fahrenheit in center and cooler at each end; a higher temperature would have greatly assisted classification.

Double effects concentrated the clarified juice to 40° Brix, and the strike pan completed the work.

Although the semi-sirup contained a purity often above 70, it was difficult and generally impossible to start a grain in the pan; a strike thus boiled to grain produced exceedingly fine grain, difficult to purge and invariably dark in color, no better than a number of early strikes boiled to string. These fine, gummy, dark sugars, dissolved in clarified juices, were used to start the grain; an amount equal in weight to one-fifth that of each strike produced a fine sugar of medium size grain, remarkable for its uniformity of grain, color, and purity. All sugars were taken to the mixer and passed through the contrifugals as speedily as possible to remove them from contact with the black molasses.

The entire water supply was obtained from a bed of gypsum 65 feet from the surface, and was positively unfit for use in either the boilers or the diffusion battery. The injurious effects of this water were observed early, Dr. Wiley being the first to suspect the true cause. By the use of this water for diffusion there is a loss (estimated) of 22½ pounds of sugar from each ton of cane worked, or 35 per cent. It ruined the molasses, and to this gypsum is attributed, directly or indirectly, nearly two-thirds of the annoying and expensive delays and losses incident to the present season's work.

Canes of unusual richness were worked, the battery secured a good extraction, the entire evaporation occurred in vacuum with but slight inversion of sugar; but large yields of sugar did not follow. The analyses of molasses from the sugars explain much, many of them showing the relative sugars four and even four and one-half to one, yet so engulfed with a mass of gums black and bitter as to render impracticable any attempt to secure second sugars. In my opinion, the estimated loss of sugar due to the use of this water should be doubled. I would respectfully ask critically inclined persons to keep these facts in mind when reviewing the accompanying tables, which contain, notwithstanding, some interesting and reliable information.

The farmer looks upon this industry as one created for his especial benefit, and when considered from his stand-point as judged by its agriculture, can see only magnificent successes for all sugar work. An average crop of cane as grown in this section at $2 per ton equals in value the land upon which it is grown. No crops are grown with more certainty; others, corn especially, in most localities of this section are not sure every season. One farmer growing 30 acres reports an average yield of 13½ tons per acre. Some small pieces produced more, the average being 10½ tons per acre. Ten thousand acres of cane at $2 per ton could easily be contracted for delivery next season. The farmers are not slow to see the advantages offered in growing cane at these prices.

The soil of this section can be called neither clay nor sand, being light, loose, not sticky, light in color, contains little organic matter and produces only a medium-sized stalk of corn or cane.

I attribute the phenomenal richness of canes grown here the past season to warm soil, high elevation—1,500 feet above sea-level—pure, dry atmosphere, proper selection of seed, good culture, and long period of hot, dry weather; the latter acting to some extent as an unfavorable condition of growth, the plant in its efforts to reproduce itself developing a higher content of sugar.

Dark, heavy soils produce a stalk of abnormal size, continuing its growth until checked by frosts, containing invariably a large per cent. of reducing sugars.

Light, thin soils produce an undersized stalk perhaps 4 feet long, maturing but a handful of seed, generally showing a high per cent. of sugar and often a very low per cent. of reducing sugars. If these conclusions are correct, the elevated table-lands of southwestern Kansas, situated directly south and west of the Arkansas River Valley, will offer inducements for the prosecution of this work not found in localities north and east of that valley.

A remarkable feature of this season's crop was its high average content of sugar, low per cent. of reducing sugars, and the disposition to increase the former at the expense of the latter for nearly two months after the cane had matured its seed. The last analysis of stalk cane made November 12, from field cane twice frozen, was 13.85 per cent. sucrose, 1.01 per cent. glucose.

I believe this crop of cane the richest by far of any ever grown and worked for sugar.

But one trial run was made, worked by itself: 43 tons of cleaned cane, from which were obtained 3,850 pounds of sugar of 98 per cent. purity and 1,000 gallons of molasses, being 90 pounds of sugar and 23.2 gallons of molasses from each ton. The laboratory work under the direction of Dr. H. W. Wiley, in charge of Prof. E. A. Von Schweinitz, assisted by Mr. Oma Carr, has been most satisfactory. The information gained through their labors will prove very interesting and valuable to all friends of this industry.

I am well satisfied no well-regulated sugar works can be successfully operated and the best results obtained unless a complete chemical control of the every-day work prevails.

Their services are invaluable as a check upon the work of diffusion and clarification. A change from hard to softer cane or a slight alteration in the adjustment of the shredder may result in great loss of sugar in the former; a change in the treatment of the juice results in loss by inversion in the latter. The cause and extent in each case are disclosed only by the chemist's art.

An expenditure upon this plant of $2,000 or $3,000 for an additional boiler and cutters would give it a working capacity of fully 150 tons per day with a full equipment of new and modern machinery. This plant could now be duplicated for much less money.

To the unfriendly critic the statements herein made will be a source of comfort, for, alas, nothing succeeds like success; results, not causes, are wanted, and no mitigating circumstances or unfavorable conditions are considered. Nevertheless those best informed see much that is very encouraging.

A new and desirable system of storing and preparing cane for diffusion was tested, its advantages proven, and its weak points disclosed; this, with the high per cent. of sugar found in this cane, is a fair offset for losses sustained by weak cutters and the use of gypsum water.

The following facts may not be out of place: This enterprise was no exception to those preceding in respect to starting late in the season, after the crop was planted, as it were. Less than three months intervened between the placing of orders for the machinery and the date of ripening of the first planted cane. The factory was two weeks late in starting and the other end of the season shortened by burning of the boilers November 4, leaving 75 acres of most excellent cane that was rich in sugar.

The gypsum had a most disastrous effect upon the boilers; frequent stoppages of work were required to clean them. By reason of excessive scaling of boiler shell and tubes the efficiency of the boilers was greatly reduced.

The following figures relative to this plant were taken from the books of the company and are reliable:

Cost of sugar-works plant	$44,547.72	$44,547.72
Less cost of water-works plant	6,000.00	
	38,547.72	
Donation city water-works bonds	12,800.00	
Received from U. S. Department of Agriculture	10,000.00	
Farmers' stock, for cane, paid in	4,500.00	
		27,300.00
Cost to present owners		17,247.72
Cost of labor		5,896.02
Less labor on water-works		1,500.00
		4,396.02
Cost of fuel		3,096.33
Cost of cane		5,980.00
Cost of incidentals, barrels, etc		1,364.37
		14,836.72
100,000 pounds of sugar at 6½ cents	$6,500.00	
100,000 pounds sugar, 2 cents State bounty	2,000.00	
36,000 gallons of molasses at 12 cents	4,320.00	
6,000 bushels seed, 50 cents (estimated)	3,000.00	
		15,820.00
Gain		983.28

Five thousand dollars were paid to railroads for freight transportation. The cost for coal and labor to handle 1 ton of cane is 2.50 cents; much coal was used for testing machinery, water-works, etc. Profit per ton over cost of production, 33 cents. Taking the season as a whole the plant was operated at less than half its capacity with no decrease in cost of labor. Fully 150 tons could have been worked with the same labor and an increase of 20 per cent. of fuel, making the value per ton of cane worked over cost of production 1.62 cents, or $243 per day.

For working a 200-ton plant, costing perhaps 20 per cent. additional above this one, labor and incidentals increased 10 per cent. and fuel 25 per cent., would show value of product over cost of production of 3.60 cents per ton, or $720 per day. These yields are based upon results of this season's work, 60 pounds of sugar and 16½ gallons of molasses from each ton, which certainly is 20 per cent. less than may reasonably be expected by the use of good water.

The average quality of sugar as placed upon the market from these works was equal to the best in purity, but stained slightly by contact with black molasses. It has a hard, firm, medium sized, well-cut grain, was dried thoroughly, and unlike all fine-grained sorghum sugars heretofore produced does not cake or become hard in the barrel. It stands next to granulated in price and sweetening power, the jobber selling at 6⅞ cents per pound more of this sugar than all yellow sugars combined. Confectioners appreciate its sweetening power. The molasses was very dark in color, sharp and bitter to the taste, classed but little better than black-strap; with pure water the quality would be improved and the selling price increased to 18 or 20 cents per gallon.

Unless some means are devised for removing a larger per cent. of the impurities in the juice or a less quantity of sugar is secured, enabling the production of a molasses suitable for table use, the near future will see enormous quantities of molasses produced fit only for mixing purposes, for which the demand is necessarily limited.

A plant working 200 tons per day will produce annually 250,000 gallons of molasses, and unless suitable for table use it must be used for fattening hogs and cattle or converted into alcohol.

The Department of Agriculture, under the direction of Dr. H. W. Wiley, who first advocated and practically applied the process of diffusion to the manufacture of sugar from sorghum, has made it possible to secure practically all the sugar in the juice, this being the first and greatest step toward the establishment of the industry; the next greatest and scarcely less important step still awaits a solution. I refer to the clarification of sorghum juices. The methods now employed for this purpose are borrowed from the sugar-cane work of Louisiana, being merely the addition of lime and removing what scums appear on the surface.

Analysis shows the amount of sugar in each ton of cane, averaging the whole season, to be 249 pounds; the glucose would hold in solution 66 pounds, leaving 183 pounds available, did not other solids, as gums, starch, coloring matter, etc., also restrain ¼ times their equal of sugar from graining until a possible yield of 100 pounds or less from each ton of cane is our best work. Must we stop here and permit the loss of one-half or more of the sugar found in the cane? The task is not an easy one as the many know who have considered it even briefly, but its importance and necessity demand that we sit not idly by.

The people of the whole southwestern portion of this State to my personal knowledge are enthusiastic upon the question of sorghum sugar; a failure any season to grow good sorghum is not recorded. The establishing of sugar works would bring under cultivation lands now considered of little value except for growing sorghum, and fortunately will produce a sorghum of the very best quality for producing sugar.

These facts are fully appreciated, and every town, many without water, and others without railroads, aspires to the possession of a sugar works.

Daily during the working season committees, delegations, and individuals visited the sugar works, leaving full of confidence in the work.

A number of factories could be erected in this section next season if experienced men could be found to operate them.

ABSTRACT OF MR. DEMING'S REPORT TO THE CONWAY SPRINGS SUGAR COMPANY.

To President and Stockholders of Conway Springs Sugar Company:

I hereby submit for your consideration the following report of your works the past season:

I would especially call your attention to the following facts: A complete organization was not effected until about April 20. Orders for machinery were placed about June 1; very little machinery had arrived July 1; all the heavy machinery was on the ground July 25, the strike pan and boilers only being placed. Boilers were first fired August 15; cane-shed and cutters tested August 22; first chips taken to battery August 26. On this latter date was completed drags and an arrangement for drying chips to be used as fuel. The two following days they were tested, and removed during the next two. Eleven cells of chips were diffused August 28 and 29 and concentrated in the strike-pan. Regular work began September 1 on early cane, producing only molasses. September 12 began work on orange cane for sugar.

From the foregoing you will observe the late date of organization, the necessarily short time for selecting suitable machinery, and also for its manufacture. From special designs most of it was manufactured in the East at increased cost. This, in connection with the time and labor required for placing the same and making the necessary pipe connections throughout the building with a class of mechanics and laborers without previous experience with this line of machinery, accounts for starting the factory two weeks after the cane was ready.

For growing cane the season was unfavorable. Sod cane and late planting were greatly injured by the drought.

Fifty farmers contracted to grow 600 acres of cane. One hundred and five acres of old ground were planted with amber seed represented as pure, but badly mixed with orange, which was worked green, contained no sugar, and too immature to produce good molasses. Fifty-four acres produced 8 tons per acre; 51 acres remain unworked, of no value except as fodder. One hundred acres of sod were planted to orange; 50 acres produced 5 tons per acre; the other 50, planted late, is only suitable for fodder. Three hundred and ninety-five acres of old ground were planted to orange; 220 acre were worked, producing 10½ tons per acre; 175 acres remain unworked; of this, 100 acres, late planting, only fit for fodder, while 75 acres of most excellent cane yet remain in the field.

A few acres of Link's Hybrid variety of cane were grown, making a satisfactory growth, but inferior to the Orange both in sugar content and power of retaining its sugar after the seed had matured.

A slight frost occurred October 25, and a heavier one November 4, doing no damage. A freeze occurred November 9, and again on the 11th.

The unfortunate burning of the boilers November 4, when machinery was working well, with cane in its best condition, and the prospect good for working the whole crop, is indeed to be regretted.

The farmers are to be congratulated on their readiness to grow cane, disposition to aid the enterprise by taking stock, paying therefor in cane, and their success in producing a crop of cane never before equaled in its sugar-making properties.

A very remarkable fact developed by the factory work was the canes' unusually high content of sugar and its disposition to not only maintain but increase its sugar content, at the same time decreasing its invert sugar. The fact that this cane had matured its seed nearly two months previous, some having been frozen and thawed twice by November 12, date of last analysis, indicates that this section, by its high elevation, dry atmosphere, absence of early frosts, and peculiar soil, has, so far as my knowledge extends, advantages not possessed by other localities.

The last analysis of field cane was made November 12, 13.85 sucrose and 1.01 glucose. August 15 cane was in condition for making sugar, and remained so until November 15, providing a three months' working season, nearly one month longer than at Fort Scott. Sixty-three analyses of cane chips, fully representative of the crop and the season, averaged 12.45 per cent. of sucrose (true sugar), and 2.37 per cent. of glucose (reducing sugars.) The average of fifty-three analyses taken at Fort Scott last season was 9.54 sucrose, and 3.40 glucose. Admitting these juices contained no other solids not sugar, except the glucose (which is not true), yet granting an equal per cent., the cane grown here has 182.8 pounds available sugar per ton, against 95.6 pounds at Fort Scott.

As further evidence of the phenomenal conditions prevailing here I would call your attention to the averages of analyses from which the above was taken:

No.	Dates.	Sucrose.	Glucose.
24	Analyses September 12 to 30	11.15	2.92
35	Analyses October 2 to 30	13.23	2.07
4	Analyses November 1 to 4	13.45	1.69
63	Total........	12.45	2.37

Note the increase of sucrose and the corresponding decrease of glucose. Such relations of the two sugars in sorghum, existing for a period of two months, are without precedent in the whole history of the industry, and suggest that possibly the area over which this business may be conducted with the greatest success is not limitless, as some suppose.

The cane worked produced about 6,000 bushels of seed; 2,163 tons of cleaned cane were worked for molasses, producing 36,000 gallons, or 16.6 gallons per ton, in addition to the sugar; 1,673 tons of cleaned cane were worked for sugar, producing 100,000 pounds, or 60 pounds per ton; 240 tons were lost—in the fans, 3; not drawn from battery, 117; soured in battery, 20; soured semi-sirup, 40; left as semi-sirup when work ended, 60.

A supply of water sufficient for the water-works also was obtained at considerable cost.

The well furnishing the supply being 15 inches in diameter and 50 feet deep through a substance known as keel, an 8-inch drill-hole was carried 15 feet below; into this was placed a 5-inch suction-pipe.

The maximum supply of water was equal to the discharge, under slight pressure, of a 3-inch pipe, inadequate for factory work; 500 feet distant was formed a pond from which a 3-inch discharge pump supplied the boilers, diffusion battery, and double effects; two days' time were lost in making this change, but an adequate supply of water was obtained.

I have endeavored to prepare a statement whereby the expenses incurred in operation could be shown and the real profit or loss upon this season's work clearly shown. Much labor belonging to plant account has been charged to operation. A complete separation of the water-works and sugar company's interests in this respect is impossible. Quite a quantity of fuel is on hand, some work yet remains to be done, and further, very much of the product remains on hand unsold. Therefore, any statement now offered will in great part be assumed. However, with figures now at hand, and estimating value of products from prices already obtained, I may confidently assert that the product of this season's work considerably more than equals in value the cost of its production. This is a very creditable showing indeed, especially when we consider that from one cause or another the factory, taking the season as a whole, averaged less than half its capacity without a corresponding reduction in the operating expenses.

MR. DEMING'S DIRECTIONS FOR RAISING CANE.

Much depends on a good stand from the first planting. No filling-in will be allowed. If necessary to replant any portion, it must be replowed, cultivated, or listed over.

The field should first be cleared of all trash, such as stalks, weeds, and bunches of grass. This is best done by raking and burning. Unless a lister is used a good seed bed, such as for wheat, should be provided, and the seed deposited in fresh, moist earth, deep enough to insure moisture, yet not beyond the sun's warmth. This varies from one-half inch in depth on heavy clay soils to 3 or more inches on light, loose, sandy soil.

It is essential that the seed be planted at an even, uniform depth to insure its coming up and ripening early, and the seed must under no circumstances be dropped or covered by hand. For loose sandy soils a lister is a good planter. A good garden drill may answer, and under some circumstances a forced wheat drill, having all the holes except the two next the outside ones closed; but for a prepared seed bed a regular two-horse corn planter, with or without a drill attachment, gives the best results, planting at a uniform depth, and the wheel firming the soil about the seed, causing it to germinate and grow more rapidly, with a better start of the weeds.

Unless the planter has broom-corn plates, which are the best, the holes in the corn plates should be partially closed, with lead, babbitt, cork, or leather, until they admit of the passage of not more than four or five seeds at each movement of the plate. A slight excess of seed should be planted, and the hoe used to properly clean it out. This should be done invariably before the cane is 4 inches high. Good soils will produce a stalk of cane for each 4 inches of row space. When the rows are 42 inches apart, two stalks should be allowed a space of 10 inches, three stalks 18 inches, four stalks 30 inches, six stalks 42 inches, and never more than six stalks in any one bunch, no matter how spaced.

Foul land is easiest tended when planted in checks, and all lands so planted produced more sugar, but a smaller tonnage, than when planted in drills. The cultivation should be merely upon the surface to avoid cutting and otherwise disturbing the roots, checking their growth, and inducing a growth of suckers to sap the parent stalks and retard their development.

All that is required is to keep the grass and weeds in check, and all cultivation should cease when the joints appear, as any interference with the roots at this time results most seriously. One well-matured stalk will grow on the space occupied by two small ones, is as heavy as six small ones, and contains more juice sugar and less impurities in proportion to its weight. The seed and leaves are less than 25 per cent. of total weight of the large stalks, while with small canes the loss from this source may reach fully 50 per cent.

To plant cane upon new ground the turned sod should be quite thin, but evenly and smoothly laid. The seed should be planted with a two-horse corn-planter, provided with a rolling coulter to cut and not displace the sod, depositing the seed just underneath the subsoil. The sod acts as an excellent mulch to retain moisture and prevent the growth of grass and weeds, no cultivation or further attention except thinning being necessary until harvest time.

A good practice for planting cane upon old ground is to plow the land at any time during early spring, but do not harrow. At planting time take a two-horse cultivator, place three small shovels upon each beam, spread and fasten the beams so that the shovels will work up a space for two rows each 4 inches deep and 12 inches wide. Let the planter follow soon, depositing the seed in the center of this worked-over space. There will be no weeds or grass for 6 inches upon either side of the plants, and the cultivator will care for the space between the rows. Cane deteriorates very rapidly when cut, lying on the ground in bunches, exposed to the sun and drying winds, a few days of such exposure changing the sugar into glucose.

Cane should be delivered the same day as cut, the only exception to this rule being to cut and load on the wagon the evening before what can be delivered early the next morning.

Next to the importance of properly thinning the canes the necessity of having well-matured, freshly cut, promptly delivered cane is the most important point connected with the agriculture of this business.

Instructions for converting an ordinary hay-rack into a cane-rack will be furnished by the cane agent. Each wagon must be provided with two ropes, each three-fourths of an inch in diameter and 35 feet long, by which the cane is unloaded. The cane must be loaded so the tops project over the right side of the rack, facing the team.

REPORT OF E. A. v. SCHWEINITZ.

The character of the growing season of 1888, for sorghum, in the vicinity of Conway Springs, record of which was kept by Mr. J. M. Wilson, the cane-grower, was the following:

From April 16 to 21, when the first planting was made, the ground was still cold, but otherwise in good condition. April 21 to 25 the weather was cool and cloudy, followed by heavy rains on the 26th and 27th, and by heavy frost on April 30, which froze the ground one-half inch.

The beginning of May was clear and cool, with rain on the 6th, followed by clear and warmer weather up to the 16th, with rain on the 17th, warmer weather until the 24th, when there was again a heavy rain. The month of June was warm, with good rains upon the 8th, 21st, and 26th.

July and August were exceedingly hot months, with scorching winds, but with a good rain on July 14, and light rain on August 5 and 6.

September and October were hot and dry, with no rain until October 21. The first heavy frost occurred November 4, but did not damage the cane. The first freeze was on November 9. Already on October 25 there was light frost, but not sufficient to kill the leaves, and by November 2 they were thoroughly dry and dead.

Work stopped on November 4, and November 8 there was a heavy snow-storm and blizzard. The last cane was analyzed a week after the factory stopped, but appeared as good as at any time during the season and did not at that date show any effects of the thaw following the freeze. This was due, probably, to the fact that the cane was very dry.

The elevation of Conway Springs is about 1,500 feet above the sea-level. The soil upon which the sorghum was planted is an upland sandy loam. About one-fourth of the crop was upon sod land and the rest on old plowed land.

The subsoil is derived from the decomposition of friable red shale, which contains a fairly large percentage of phosphoric acid with but little potash.

The first planting was Early Amber seed, supposed to be pure, and the later planting of Orange, Sterling Orange, and on May 16 about 10 acres of Link's Hybrid.

The seed was put in either with a planter or strewn on top and harrowed. The average depth of planting was 2 inches. It was found necessary early in May to replant some of the Amber which had been covered to a depth of 4 inches. Owing to the late date at which the building was begun and the machinery ordered, the factory was not ready for work until September 1. The first cane was cut August 24, and regular work begun September 6.

The seed planted for pure hand-picked Early Amber proved to have been a mixture of Amber and Sterling Orange. In consequence, when the Amber was ready to be cut and worked, the Orange mixed with it was still green, showing a low content of sucrose. After a few days' work it was decided to have the farmers pick out the best of the Early Amber in the field and condemn the remainder. As fully one half of the crop of the supposed Early Amber proved to be this Sterling Orange, the first work of the season was of but little value, and no attempt was made to obtain sugar, all of the cane being worked for molasses.

September 4 the first Early Orange cane was cut. At that time, although the seed was hard and ripe, the content of sugar in the cane was not nearly as high as the same variety of cane showed later in the season.

The Sterling Orange was at its best about October 1, and the Link's Hybrid at the time it was worked, November 1.

The results of analyses of whole canes are recorded in Table No. 1. The canes were topped and stripped, and the juice expressed by means of a small hand-mill. The average amount of sucrose in the juice was about 2.09 per cent. higher than the average of any crop heretofore worked. The highest percentage was found in sample No. 162, taken from a load of Sterling Orange. The lowest percentage of sucrose was noted in two samples of mixed Amber and unripe Orange, on September 4 and September 10. The best samples taken during the working season were Nos. 27 Amber, 352 Orange, and 374 Link's Hybrid. The Amber cane after being cut, if left lying for any length of time, deteriorated rapidly, as shown by the analysis of No. 26.

The percentage of moisture in the cane during the month of October decreased rapidly, and the same quantity by weight of cane yielded only about one-half the weight of juice given earlier in the season. The dryness of the cane was also noted by the farmers, as their loads lost several hundred pounds as compared with the same sized load during the first part of the work. It may also be noted that the cane was very pithy. On an average, one out of every five stalks contained little or no juice and a large amount of fiber. The cane cut during October, a great quantity of which was left lying from two to three days at a time, on account of delays in working, did not deteriorate to any great extent. The dryness of the cane again probably explains this.

After the factory stopped, a number of samples of cane was taken for the purpose of determining the condition of the still outstanding crop.

Samples Nos. 382 and 388 gave the highest result of the season. Another sample, No. 383, from a field which the cane-grower claimed was the poorest out, showed a high percentage. No. 378 was from a field of second growth, from stubble. On November 4 some 25 tons of cane were left on the rack. One lot was selected and analyzed, some of it put into a silo. A sample of the remainder, tested four days later, showed that there had been no deterioration in the cane, as can be seen from analyses Nos. 386 and 391. This cane had been exposed to heavy frost, snow, and thaw.

Cane taken from the field on November 7, and again from the same field November 12, showed but little deterioration.

The average percentage of sucrose in the mill juices from the fresh chips is .3 per cent. higher than that recorded in the average of the whole canes. This is explained by noting several very low percentages of sucrose in some of the samples of whole cane, without a corresponding low percentage in the chips.

Here it may be noted that in taking samples of fresh and exhausted chips, as also of diffusion and clarified juices, care was taken to secure comparative samples. The battery consisted of sixteen cells, but only twelve of these were in the circuit at one time. The fresh chips were taken from these twelve cells and the exhausted chips from the same. The juices were sampled as they ran into the defecators, care being taken to secure those corresponding to the fresh chips. The samples of semi sirup were taken as a rule once every twelve hours, and correspond approximately to the juices analyzed. For the most part two sets of samples were taken, one in the morning and the other in the afternoon.

The lowest sucrose and highest glucose were recorded at the beginning of the season. The highest sucrose of the season was noted on October 15, and lowest glucose on October 26.

The average percentage of sucrose for October was 13.22 and glucose 2.07. From September 26 to the end of the season the mill juices appeared to be unusually rich. The average for October was .8 per cent.

higher than the average for the entire season. This is 2.88 per cent. higher than the average at Fort Scott in 1887. As noted in connection with the whole canes, the dryness may partly explain this, but the location and soil of Conway Springs seem to be especially adapted to the growth of sorghum. It is further south than any other point in Kansas, where sorghum has been grown and the season appears to be longer and better than in eastern Kansas.

The mean of sucrose in diffusion juices is higher than the mean at Fort Scott in 1887, but considerably lower than would be expected from the analyses of the chip juices. The difference may be accounted for either by the dryness and pithiness of the canes, as just mentioned, or by inversion in the battery. In order to prevent inversion, if any, carbonate of lime was used in the battery for a time. Although the acid was neutralized to about the same extent as at Fort Scott, apparently inversion was not prevented. The results of the analyses are given in Tables 12, 13, and 14. In place of carbonate of lime a number of experiments were made with caustic lime. The lime was distributed upon the chips as they passed from the macerator to the battery, by means of a roll, about $1\frac{1}{4}$ pounds of lime being added to each cell.

The object was to add just so much lime to the chips that 100 c. c. of the juice when in the clarifiers would require about 5 c. c. of $\frac{n}{10}$ alkali to neutralize it. To attain this exact point was difficult and the tables in which results of the work are given show all possible variations. The lime as sprinkled on the chips also neutralized the acid in the mill juice, as may be seen from the table.

	Glucose.	Sucrose.
In the mill juices treated with lime the proportions were	1	6.6
In mill juices, etc	1	5.0
In diffusion juices:		
In juices without lime	1	4.9
In diffusion juices	1	3.3

If we note samples 183 and 184 on October 5, there appears to have been no inversion whatever. On several other days the apparent inversion was but slight. It may be mentioned further that on those days on which little or no inversion was noted, the percentage of glucose in the mill juice was high, and the amount of juice given by the cane as taken from records of weight of juice was above the average for the season. The average number of cubic centimeters $\frac{n}{10}$ alkali required to neutralize the acid in the juices was 40.6.

Solids in mill juice 19.39
Solids in diffusion juice 12.99

$$19.39 : 12.99 :: 40.6 : x.$$

Normal acidity of the diffusion juice 27.2

That is considerably higher than the acidity of juices found at Fort Scott, average of which was 19.98.

The highest per cent. of sucrose for the season in the diffusion juice was noted September 29, 10.02 per cent., being 2.30 per cent. above the average. The corresponding mill juice for the same date was 14.92 per cent. sucrose, 2.5 above the average, showing that fair comparative samples had been secured.

The average during October was 8.59 per cent. sucrose, 1.74 per cent. glucose, better than the results obtained at Lawrence, La., bearing in mind the fact that the sugar cane has less glucose. The purity of the diffusion juices was lower than that of the mill juices from the chips. This is due probably to inversion in the battery.

The column headed "extraction" in table 12 is given for the purpose of noting to what extent, if any, the extraction was diminished by the use of lime in the battery. If we compare the several instances of especially low extraction, Nos. 246, 291, and 361, with the corresponding acidity, we will note that either lime was in excess or the percentage of sucrose for the day was high, without a corresponding change having been made in working the battery, and in the amount drawn off. October 15,

In mill juice there were ....:......8.4 parts glucose to 100 parts sucrose
In diffusion juice there were ....16.5 parts glucose to 100 parts sucrose

October 5 and 25, with acidity 9, the proportions in the two juices corresponded closely.

The table shows, then, that the lime, unless in excess, did not interfere with the extraction.

Record was kept during the entire season of the amounts of sucrose and glucose left in the chips. The highest percentage of sucrose in the mill juices from these was noted at the end of the season, November 2, being 2.91 per cent. The average extraction for the entire season was 88.72 per cent. of the sugar in the cane. This is a poor extraction, being fully 4.1 per cent. lower than the extraction at Fort Scott in 1887. The average dilution for the season was 11.55 per cent. From the first of the season to October 15, 160 gallons were drawn off each time. From that date till the close of the season 180 gallons. Each cell held 1,400 pounds chips. Deducting 10 per cent. for fiber, we have 1,260 pounds juice in each cell.

Average weight of juice drawn from first of season to October 15 .... pounds.. 1,349
From then till close of season...do.:... 1,513
Mean Brix from September 6 to October 15:
    In mill juices... 18.93
    In diffusion juices... 13.05
October 15 to November 2:
    Mill juices.. 20.10
    Diffusion juices.. 12.55
Dilution from September 6 to October 15.........................per cent.. 6.50
Dilution from October 15 to close of season...........................do.... 16.60

The poor extraction was due partly to the large chips furnished by the small cutters during a portion of the season, to the irregularity in working, but chiefly to the small quantity of juice drawn off; all points which might have been more carefully noted and the loss avoided. As the dilution, if moderate, is of small importance, the object should be to get all or as nearly all as possible of the sugar from the cane.

The purity of the defecated juices, Table IV, is 1.5 points higher than the diffusion juices, due to a little destruction of glucose in the clarifiers. The table shows in addition that there was no inversion in the clarifiers. The scum from the defecators was, as is usual, about as rich in sugar as the juices themselves. These scums were thrown into the ditch, thus entailing a loss of sugar which could and should have been avoided by returning them to the battery.

For the purpose of comparing the readings of the Brix spindle with the actual total solids obtained by drying and weighing, a number of determinations were made.

Rectangular flat platinum dishes three-eighths inch in depth were used and the samples dried for five hours at 100° C. The samples were weighed from a tared flask, about 2 grams being taken in each instance. Duplicates were always made. The use of asbestus as an absorbent agent in drying was also tested. A thin layer of loose asbestus was placed in the bottom of the dish, and the sample dried at and for the same length of time as those samples where the dish alone was used.

The average of these results in the case of mill juices gives the solids 1.46 per cent. less, and with asbestus as 1.66 per cent. less, than the average of the spindle readings. These results are fully 1 per cent. lower than those recorded by Dr. Crampton at Fort Scott in 1887, but agree closely with results obtained at Douglas, Kans., and at the Department this year. In the case of the diffusion juices, the dish alone gave 1.25 per cent. solids less, and dish with asbestus 1.40 per cent. less, than the direct readings of the spindle. Correcting the percentage of sugar on this basis, in the mill juices it would be .11 per cent. higher, and in the diffusion juices .05 per cent. The purity is also largely increased by calculating on the weight of actual solids.

It may again be remarked here, as was stated in 1887, the use of the Brix spindles standardized for pure solutions of sugar give misleading results, and the solids as determined by direct drying should be relied on. As might be expected, the samples where asbestus was used gave results slightly lower than those without, and it may be stated further that the duplicates with asbestus agreed more closely. The asbestus furnishes a larger drying surface, and less moisture is retained than is the case with a thick film of the sirup. The asbestus should, then, be preferred to the plain dish.

The average ratio of glucose to sucrose in the semi-sirups is slightly higher than that in the defecated juices:

	Glucose.	Sucrose.
Semi-sirups	1	3.86
Defecated juices	1	3.68

This difference is due either to error of experiment or to the equalizing effect of large quantities of juice or to a slight inversion in the double-effect vacuum-pan.

It is interesting to note this, as it is the first time that the Lillie patent has ever been used as a double effect. It is true, the pan gave a great deal of trouble and caused a great deal of delay during the working season. This was due, first, to the fact that the pumps put in to keep up the circulation of the juice, viz, rotary pumps, were not suited to the work, and secondly and chiefly, because the effect had been hurriedly and carelessly put up by inexperienced workmen. At the close of the season centrifugal pumps were substituted for the rotaries and the pan thoroughly overhauled by an engineer from Philadelphia, and it then gave satisfaction. The inside of the juice-tubes became rapidly coated with a hard scale, which necessitated their being cleaned every four or five days. This scale was due principally to the mineral water, which will be referred to in another connection.

The first sugar made was grained in the wagons. The grain was small, and as it had been allowed to stand for a considerable length of time and become cold, it was difficult to free it in the centrifugals from the gummy matters. Table No. 7 gives the polarization of this product. All of it was reboiled and used for enriching the semi-sirups, hence the high purity and increased percentage of sucrose in the masse cuites and molasses over and above the semi-sirups. The samples of the masse cuites were taken from the mixer, and the samples of molasses taken from time to time from the storage tanks and barrels give the average composition of this product for the season. The purity of the molasses from the enriched sirups is higher than the purity of the semi-sirups at Fort Scott in 1887, but as the molasses could be disposed of, it was considered more profitable to sell it than to work it for seconds.

The water from the well proved upon examination to be highly charged with mineral matter, containing 318 grains to the gallon. This was chiefly gypsum, together with some little magnesium sulphate and sodium chloride. A 10 per cent. solution of sugar prepared with this water and evaporated to a thick sirup showed no more inversion than a solution of the same strength made up with distilled water and evaporated. The addition of acetate of lime to the solution had no inverting action.

The water gave particular trouble in the boilers, forming rapidly a heavy scale. The want of proper cleaning in the early part of the season caused burning of the boilers on November 4 and stopped the work. The latter part of the season the vapor water was run into a pond and used for diffusion purposes. This water was strongly acid, due to the decomposition of organic matter, and not much of an improvement on the well water. On account of the foaming it was difficult also to use it in the boilers.

In the few samples of masse-cuite not enriched the proportion of sucrose to glucose was about the same as in the semi-sirups, showing that there was not any inversion in the strike-pan.

The percentage of ash found in the masse cuite is 1.5 per cent. and in the molasses 1 per cent. higher than the average found at Fort Scott in 1887. This we may fairly attribute to the large amount of gypsum in the water. After pond water was substituted for the well water, except on one or two days when lime in the battery was in excess, the corresponding percentage of ash was diminished.

The indirect readings of sucrose are either lower or correspond closely with the direct readings. This points to a large amount of starch in the juices, which will further explain the high readings in the mill juices and apparent inversion in the battery. The solids not sugar are also higher than heretofore noted; also to be attributed to starchy and gummy matter.

The proportion of glucose to sucrose in the molasses is about 1 : 3. This high ratio is doubtless due to mineral and organic matters (not sugar) present, which, while not causing inversion, prevent crystallization. Further, the sugar-boiler was troubled with gummy matter in the strike-pan, and the masse cuite was very gummy in the centrifugals. On several occasions quite a quantity of white gummy matter was taken out by the proof-stick. This fully explains why such a large proportion of sucrose was left in the molasses. The analysis of sample 122 shows proportion of glucose to sucrose 1 : 1.6. This sample was from the first lot of mixed cane worked, which contained but little sucrose.

The percentage of albuminoids in the juices from fresh chips is high. The per cent. in diffusion juice is .25 lower, and in clarified juices .01 less than in diffusion, showing that the defecation had removed a comparatively small proportion of the albuminoids.

The figures show further that the cane contained an unusually large amount of fiber and organic matter not sugars, which went into the diffusion juice. This further accounts for the large amount of gummy matter in the strike pan, and, together with the large amount of mineral matter, explains the low yield of sugar from the rich cane.

The color of the sugars was a grayish yellow, and rated on the market as a little better than C brown. The grain was large and firm.

The mean polarization of raw gummy sugars was 82.52, and of first sugars 96.80.

The following is the record of the number of tons of cane worked, sugar and molasses made:

Total number of tons of cane passed over the scales 2,991

Of this 430.5 tons were Early Amber, mixed with unripe Sterling Orange.

Two thousand five hundred and sixty and five-tenths tons were chiefly Orange, with a small quantity of Link's Hybrid. The estimated average tonnage per acre is 10; the highest tonnage, 13.5 per acre. Twenty-five tons were left on the cane rack when work stopped, so that the actual number of tons of cane worked was 2,966, and tons worked for sugar 2,535.5. Tons of cane for molasses only, 430.5. Deducting 25 per cent. for leaves and seed, we have 2,225 tons of cleaned cane.

Total number of cells filled from September 12 to close...... 2,730
Number of pounds of chips in each cell.................................... 1,400
Total number of pounds of chips in cells (1,860 tons).................... 3,722,000
Number of. tons of cleaned cane from September 12 to close ....:..... 1,901

Making a difference of 41 tons unaccounted for, some of which was thrown out by the fan and from the drag. The remainder can be attributed to lost records, which were missing for several days' work.

Yield of sugar.

Total number of pounds of sugar 100,500
Gallons of molasses ....:............................... 36,000

There was left on hand at close of season one tankful of semi sirup, equal to 600 gallons of molasses. This makes average yield of sugar per ton of field cane, estimated on the cane actually worked for sugar, 39.2 pounds, and on cleaned cane 52.8 pounds. The quantity of molasses made per ton of cleaned cane was 14 gallons, or, estimating the sugar on total number of tons of cane cut during the season, we have 45.1 pounds per ton of cleaned cane. Two trial runs were made during the season. The first 46.9 tons gave 3,986.5 pounds sugar and 9,580 gallons molasses, equal to 85 pounds sugar and 20 gallons molasses per ton. The second trial run gave 90 pounds sugar and 16 gallons of molasses per ton on a run of 60 tons.

During the season there were lost by carelessness 4,800 gallons of semi-sirup and 7,200 gallons of juice, corresponding to about 100 tons of cane. The battery soured twice and was drawn off twenty-eight times, causing a loss of 192 cells of chips of 1,400 pounds each, equal to 134 tons of cane. Deducting, then, 234 tons from the number of tons worked for sugar, we have 1,667 tons of cleaned cane, with an average of 60.2 pounds sugar per ton.

From each ton it was estimated that 2 bushels of seed and 200 pounds of leaves were obtained. The seed was carefully hand-picked and thrashed, so that this product will prove very valuable.

The total number of days' actual work, counting each day at twenty-two hours, was thirty. By that we mean that the number of hours during which the cutter actually worked would be equal to thirty days of twenty-two hours each. If a factory is substantially built, the machinery strong and every bolt in its place, there is no reason why there should not be a steady yearly run of ninety days, full time. During the working season every hour's delay is so much money lost, and a sugar factory should run as smoothly as a grist mill. It is a question of practical mechanics, which a good machinist can handle.

With a number of changes in the factory the Conway Springs mill can be made a success. The foundations of the heavy machinery should be replaced; the battery put in straight line and elevated, so that the chips could be removed by means of a car; two large cutters should be ready for use in place of one; also two macerators. Three additional boilers are needed, another engine, two more clarifiers, a large storage tank for the molasses, and more sugar wagons and storage room, and, above all, a good water supply, which may with care be secured in the neighborhood.

With these improvements, with cane of the same quality as was worked this year, and careful management, a great success may be secured at Conway Springs during another season.

TABLE I.—*Mill juices from whole canes.*

Date.	No.	Total solids by Brix at 15, 5°.	Baumé.	Specific gravity.	Sucrose.	Glucose.	Purity.
		Per cent.			*Per cent*	*Per cent.*	
September 4 ..	3	15.38	8.50	1.0630	3.62	....	23.53
September 4 ..	4	18.50	10.10	1.0706	13.95	....	7?.40
September 4 .	5	16.10	8.00	1.0656	9.50	....	19.30
September 4 .	6	15.50	8.60	1.0634	8.47	....	51.64
September 6 .	9	18.91	10.50	1.0783	10.48	4.72	55.33
September 8 ..	21	18.21	10.10	1.0753	12.30	3.21	67.43
September 8	22	16.31	9.00	1.0669	10.30	3.20	63.15
September 10	24	18.91	10.50	1.0783	5.00	6.00	26.41
September 10 .	25	16.07	8.90	1.0656	11.73	3.35	72.99
September 10 .	26	21.97	12.20	1.0918	11.43	7.88	52.02
September 10 .	27	18.62	10.30	1.0770	14.32	1.28	76.90
September 10 .	28	17.21	9.50	1.0709	8.98	4.19	52.17
September 11 .	33	17.82	9.90	1.0735	11.90	2.31	66.78
September 12 .	34	18.44	10.20	1.0761	11.44	3.12	62.03
September 11	48	16.50	9.10	1.0678	11.54	2.27	69.94
September 16 .	50	17.40	9.60	1.0717	12.67	2.16	71.63
September 17 .	61	18.47	10 20	1.0761	10.61	5.51	56.89
September 18 .	69	18.81	10.40	1.0779	11.09	1.26	74.90
September 20 .	86	18.24	10.10	1.0753	10.80	3.43	59.21
September 23	114	18.51	10.20	1.0766	13.69	1.39	73.84
September 27 .	133	18.53	10.22	1.0766	12.50	1.58	67.38
October 1....	162	22.16	12.30	1.0932	16.67	2.16	75.22
October 1	163	18.85	10.60	1.0781	13.28	2.79	70.45
October 4 ...	178	21.88	12.10	1.0918	13.55	3.30	61.92
October 10....	226	19.12	10.60	1.0792	13.70	2.00	71.65
October 10....	227	17.82	9.90	1.0735	13.10	1.12	73.51
October 16 ...	261	20.24	11.20	1.0841	13.98	2.88	69.07
October 20....	281	20.40	11.30	1.0850	13.63	1.91	66.81
October 31 ...	352	20.66	11.40	1.0862	14.58	1.29	70.71
November 2 ..	358	19.29	10.70	1.0801	12.68	1.43	65.73
November 2...	367	15.22	8.40	1.0 21	7.26	2.13	47.70
November 3 ..	369	13.30	7.40	1.0540	7.51	1.54	56.46
November 3 ...	374	20.00	8.20	1.0832	14.81	.89	74.50
November 5...	375	17.63	9.75	1.0726	12.87	1.65	73.00
November 5...	378	20.36	11.30	1.0850	14 09	2.14	72.15
November 5...	379	15.10	8.40	1.0617	8.54	2.35	56.55
November 5...	380	20.20	11.20	1.0841	15.26	1.15	75.64
November 5...	382	20.00	11.10	1.0832	14.95	.77	74.75
November 5...	383	20.20	11.20	1.0841	14.85	1.21	73.51
November 6...	384	19.60	10.85	1.0815	14.25	1.51	72.70
November 7...	385	16.67	9.25	1.0687	11.80	1.67	71.32
November 7...	386	17.67	9.80	1.0730	12.38	1.69	70.06
November 7...	387	17.87	9.90	1.0739	13.15	1.00	74.71
November 7...	388	19.27	10.70	1.0801	15.11	.59	78.41
November 10..	391	17.99	10.00	1.0744	12.80	1.55	71.15
November 12	392	18.17	10.13	1.0753	13.85	1.01	76.22
Maxima ..	.....	22.16	12.30	1.0932	16.67	7.88	78.41
Means....	.....	18.35	10.10	1.0759	12.14	2.35	65.74
Minima...	....	13.30	7.40	1.0540	8.47	.77	23.53

Description of samples of whole cane

3. Amber and unripe Sterling Orange mixed.
4. Early Amber from cane-shed.
5. Unripe Sterling Orange, selected stocks.
6. Unripe Sterling Orange, suckered upper joint.
9. Amber and Sterling Orange mixed, cut and lying in shed for two days.
21. Cane from shed, mixed lot.
22. Orange from field.
24. Mixed lot from shed, cut forty-eight hours; green.
25. Orange cane from field.
26. Early Amber from wagon.
27. Early Amber from field of Mr. Troeger.
28. "Southern Red" from field.
33. Early Amber, average from ten loads.
34. Cane from shed.
48. Cane from shed.
59. Orange cane from shed, cut and lying forty-eight hours.
61. Sterling Orange from wagon.
69. Early Amber from load brought in by Troeger.
86. Orange cane from shed.
114. Orange cane from field.
133. Links Hybrid from wagon.
162. Sterling Orange from wagon.
167. Sterling Orange, average from three loads.
178. Orange from shed, lying forty-eight hours.
226. Orange from shed and wagons, average lot lying forty-eight hours.

227. Links Hybrid from field.
261. Lot of badly suckered Orange cane from shed.
281. Cane from shed, lying thirty-six hours.
352. Orange cane average, late planting gave but little juice.
358. Orange cane from shed after first frost.
367. Links Hybrid from shed; red pith.
369. Links Hybrid from shed.
374. Links Hybrid from shed; good load.
375. Cane from shed, chiefly Orange, lying twenty-four hours.
378. Orange cane from field, second growth.
379. Orange cane from field, red pith.
380. Orange cane from field of J. R. Duncan; average lot left uncut.
382. Orange cane from field of J. S. Clark.
383. Orange cane from field said to be the poorest field out.
384. Orange cane from shed, put in silo November 6, covered with 2 to 3 feet ground
385. Orange cane from shed from same lot as put in silo November 6; cane from top of pile.
386. Orange cane from shed from same lot as put in silo November 6; average sample.
387. Orange cane from field of L. Berry.
388. Orange cane from field of Hanna.
391. Orange cane from same lot as put in silo, lying in shed eight days exposed to heavy frost, snow and thaw.
392. Orange cane from field of Hanna, from same plat as No. 388.

TABLE II.—*Mill juices from fresh chips.*

Date.	No.	Baumé.	Total solids by Brix at 15.5°.	Specific gravity.	Sucrose.	Glucose.	Albuminoids.	Purity.
			Per cent.		*Per cent.*	*Per cent.*	*Per cent.*	
September 6..	10	10.80	19.47	1.0810	6.36	7.16	....	32.66
September 7..	15	10.50	18.97	1.0783	8.93	6.34	....	47.07
September 7..	16	9.90	17.80	1.0766	8.01	5.07	....	53.78
September 10.	29	9.70	17.54	1.0722	9.73	6.02	....	55.41
September 12	35	10.50	18.94	1.0783	11.93	2.97	....	62.98
September 12.	36	10.50	18.95	1.0783	12.05	2.43	....	66.75
September 13.	43	9.90	17.87	1.0739	12.51	2.68	....	70.00
September 14.	49	9.75	17.50	1.0726	10.30	4.58	....	58.65
September 15	56	9.90	17.78	1.0730	10.73	3.42	....	60.54
September 17.	60	10.20	18.44	1.0761	8.03	4.36	.6563	43.65
September 18.	68	9.00	16.24	1.0665	11.22	2.10	....	60.08
September 18	70	8.90	16.14	1.0660	10.90	2.37	.6183	67.53
September 19	80	9.90	17.84	1.0735	9.68	4.57	....	54.26
September 20	87	10.10	18.19	1.0753	11.14	4.03	....	61.24
September 21	92	9.75	17.66	1.0726	9.34	4.11	....	52.88
September 21	98	9.25	16.78	1.0691	10.55	2.89	....	62.87
September 22	103	9.75	17.62	1.0726	....	3.62	....	
September 22	107	10.50	18.91	1.0783	11.49	3.11	....	60.76
September 24	111	9.50	17.21	1.0709	9.65	3.02	....	56.07
September 24	116	9.75	17.64	1.0726	10.89	2.34	....	61.73
September 26.	124	10.50	18.98	1.0788	12.74	2.51	....	67.12
September 27	128	11.45	20.73	1.0864	13.47	2.92	....	64.07
September 27	131	10.70	19.35	1.0801	12.78	2.68	....	60.61
September 28	138	10.60	19.24	1.0797	13.51	2.28	.8250	70.22
September 29	143	10.60	19.12	1.0792	12.79	2.36	.7438	66.89
September 20	147	11.30	20.50	1.0855	14.92	1.18	....	72.78
September 30	154	10.85	19.64	1.0815	13.82	1.80	....	70.31
September 30	158	10.60	19.25	1.0799	13.64	1.81	....	70.96
October 2..	165	10.85	19.59	1.0815	12.60	2.46	.7503	64.76
October 2...	171	10.90	19.68	1.0815	13.25	1.73	....	67.22
October 3....	176	10.20	18.49	1.0766	12.83	1.61	....	69.33
October 5....	179	11.30	20.50	1.0855	12.47	3.03	....	60.83
October 5....	183	11.00	19.83	1.0824	10.97	3.66	.6003	55.32
October 6....	189	11.30	20.40	1.0850	13.97	2.40	....	68.48
October 6	105	10.40	18.78	1.0777	12.56	2.02	....	66.80
October 8.....	206	10.40	18.80	1.0779	12.56	2.85	....	66.80
October 8	211	11.00	19.80	1.0824	12.03	2.31	.8375	60.75
October 9....	216	10.50	19.04	1.0788	11.97	2.94	....	62.86
October 11....	230	11.05	21.60	1.0904	13.28	2.86	....	61.48
October 11....	234	10.60	19.08	1.0792	12.45	2.44	.7313	65.24
October 12....	239	12.55	22.09	1.0954	10.75	2.77	....	47.37
October 13....	244	11.70	21.20	1.0886	10.05	3.91	....	47.40
October 15 ..	253	12.30	22.34	1.0936	15.58	1.24	1.0375	69.74
October 17....	262	12.20	22.10	1.0926	15.49	2.00	....	70.00
October 18....	268	11.70	21.20	1.0884	14.14	2.70	....	66.69
October 18...	272	11.50	20.81	1.0868	14.24	1.86	....	68.42
October 19...	276	12.05	21.81	1.0914	14.84	2.91	....	68.04
October 19...	277	12.20	22.03	1.0923	14.31	2.61	....	64.95
October 20...	282	12.05	21.80	1.0914	15.04	1.84	....	68.99
October 20....	283	11.10	20.00	1.0832	13.37	2.19	....	66.85
October 22....	287	10.20	18.40	1.0761	12.47	1.21	....	67.60
October 23....	292	9.80	17.73	1.0730	11.98	1.48	....	67.57
October 23 ..	296	9.75	17.63	1.0726	11.00	1.48	....	67.48
October 24....	304	10.60	19.07	1.0792	12.06	1.29	.4875	67.96
October 24....	308	10.20	18.51	1.0766	12.19	1.30	....	65.85
October 25....	313	11.20	20.34	1.0846	14.41	1.74	....	70.84
October 26....	317	11.70	21.23	1.0886	14.98	1.00	.8306	70.56
October 27....	322	11.40	20.56	1.0859	14.20	1.34	.8687	69.50
October 27...	326	10.35	18.66	1.0775	12.91	1.20	....	69.18
October 29...	331	11.40	20.60	1.0859	14.49	1.25	....	73.39
October 29....	335	10.70	19.37	1.0800	14.01	1.20	....	72.32
October 30....	340	11.40	20.57	1.0859	13.90	1.42	....	68.01
October 30....	349	11.30	20.37	1.0850	13.59	2.00	....	66.72
November 1 ..	353	11.60	21.43	1.0895	14.30	2.23	....	66.72
November 2 ..	350	11.40	20.64	1.0859	13.67	1.99	.9656	66.23
November 2 ..	363	10.60	19.09	1.0792	12.94	1.13	....	67.78
November 3 ..	370	10.35	18.67	1.0775	12.86	1.41	....	68.88
Highest.	....	12.55	22.60	1.0954	15.58	7.16	....	72.78
Means	....	10.72	19.39	1.0805	12.42	2.61	.7857	63.84
Lowest	....	8.00	16.14	1.0660	6.36	1.00	....	32.66

TABLE III.—*Diffusion juices.*

Date.	No.	Baumé.	Total solids by Brix at 15.5°.	Specific gravity.	Sucrose.	Glucose.	Albuminoids.	Purity.
					Per cent.	*Per cent.*	*Per cent.*	
September 3 ..	2	6.30	11.30	1.0435	4.98	....	....	44.07
September 6 ..	11	8.15	14.68	1.0600	5.52	6.11	....	37.60
September 7 ..	18	6.70	12.08	1.0485	4.90	4.30	....	40.56
September 10 .	30	7.30	13.11	1.0531	5.93	4.85	....	45.23
September 12 .	38	7.50	13.52	1.0548	8.39	2.40	....	62.05
September 12	39	6.80	12.22	1.0493	7.69	2.33	....	62.92
September 13 .	45	7.80	14.14	1.0574	8.37	3.20	....	59.26
September 14 .	52	6.05	10.90	1.0439	5.04	2.80	....	46.24
September 15 .	57	7.60	12.60	1.0510	7.23	2.61	....	57.38
September 17 .	62	7.60	13.77	1.0610	7.94	3.27	....	57.66
September 18 .	72	6.40	11.64	1.0468	7.35	2.16	.4678	63.14
September 19 .	77	6.55	11.85	1.0476	6.36	2.37	.4688	53.71
September 20 .	89	7.65	13.85	1.0561	6.90	2.50	.5438	49.00
September 21 .	94	7.30	13.13	1.0531	6.92	3.11	.5313	52.70
September 21 .	100	7.30	13.12	1.0531	7.17	3.07	....	54.64
September 22 .	101	6.40	11.48	1.0464	5.25	2.71	.4688	45.73
September 22	108	6.70	12.00	1.0485	6.65	2.20	....	55.41
September 24 .	113	6.80	12.24	1.0493	7.69	2.20	....	62.82
September 24 .	118	7.00	12.60	1.0510	8.20	1.91	.5188	65.08
September 26 .	125	7.05	12.70	1.0514	6.68	2.23	....	52.59
September 27 .	129	7.50	13.58	1.0553	7.55	2.67	....	55.59
September 27 .	135	6.90	12.42	1.0502	8.43	2.23	.5188	67.87
September 28 .	139	7.10	12.77	1.0514	7.92	2.24	.5313	62.01
September 29 .	145	7.65	13.85	1.0561	8.85	1.96	.5813	63.89
September 29 .	148	8.15	14.71	1.0600	10.02	1.68	....	68.11
September 30 .	155	7.50	13.50	1.0548	8.86	1.84	....	65.63
September 30 .	159	7.60	13.67	1.0553	9.08	1.91	....	66.42
October 2	166	7.50	13.50	1.0548	8.25	2.16	.5625	61.11
October 2	169	7.50	13.64	1.0553	8.98	1.89	....	65.87
October 2	173	7.60	13.67	1.0555	9.08	1.95	....	66.42
October 5	180	7.60	13.65	1.0555	7.75	2.59	....	56.79
October 5	184	7.70	13.94	1.0566	8.09	2.63	.5813	58.22
October 6	192	8.20	14.82	1.0604	9.43	2.19	....	63.63
October 6	197	7.40	13.38	1.0544	8.59	1.87	....	61.79
October 8	207	7.60	13.68	1.0557	8.28	2.08	....	60.52
October 8 ...	212	7.50	13.64	1.0553	8.59	1.89	.5813	62.90
October 9 ...	217	7.50	13.53	1.0550	7.61	2.19	....	56.66
October 11 ...	232	7.65	13.85	1.0564	8.14	2.59	....	58.90
October 11 ...	235	7.90	14.34	1.0583	9.11	2.22	.5813	63.52
October 12 ...	240	7.05	12.70	1.0514	6.72	2.37	....	52.91
October 13 ...	246	8.30	14.90	1.0609	9.48	2.36	....	63.62
October 15 ...	255	7.65	13.79	1.0561	9.37	1.55	.5038	67.94
October 17 ...	264	7.60	13.67	1.0553	6.17	1.64	....	45.19
October 18 ...	269	7.60	13.67	1.0555	8.29	1.56	....	60.43
October 18 ...	273	7.80	14.07	1.0574	8.47	1.97	....	60.19
October 19 ...	278	7.80	14.05	1.0572	8.53	2.13	....	60.61
October 20 ...	284	7.60	13.72	1.0557	8.98	1.51	....	65.45
October 22 ...	288	7.00	12.65	1.0512	7.93	1.19	....	62.69
October 23 ...	293	6.00	10.78	1.0434	6.93	1.27	....	64.28
October 23 ...	297	6.30	11.43	1.0459	7.35	1.25	.5281	64.34
October 24 ...	306	6.30	11.41	1.0459	7.20	1.27	.5094	63.10
October 24 ...	309	7.05	12.75	1.0514	8.13	1.24	....	63.76
October 25 ...	315	6.80	12.28	1.0497	8.64	1.19	....	70.35
October 26 ...	319	6.90	12.38	1.0502	8.58	1.15	.5500	69.30
October 27 ...	324	7.20	13.08	1.0531	8.90	1.28	.5906	68.04
October 27 ...	327	7.05	12.68	1.0514	8.42	1.29	....	66.08
October 29 ...	333	6.70	12.03	1.0485	7.42	1.15	.5000	61.67
October 29 ...	336	7.30	13.19	1.0536	8.26	1.33	....	62.62
October 30 ...	342	6.20	11.22	1.0451	7.33	1.20	....	65.32
October 30 ...	350	6.80	12.18	1.0493	7.20	1.36	....	59.11
November 1...	354	7.40	13.41	1.0544	8.13	1.40	....	63.10
November 2...	361	7.05	12.72	1.0514	7.48	1.80	.5187	58.91
November 2...	364	6.20	11.13	1.0447	6.92	.99	....	62.17
November 3...	371	6.90	12.47	1.0506	8.33	1.30	....	66.80
Maxima ..	....	8.30	14.82	1.0604	10.02	6.14	.5938	70.35
Means ...	....	7.20	12.99	1.0527	7.77	2.13	.5364	59.84
Minima...	....	6.00	10.78	1.0134	4.90	.99	.4688	37.66

TABLE IV.—*Clarified juices.*

Date.	No.	Baumè.	Total solids by Brix at 15.5°.	Specific gravity.	Sucrose.	Glucose.	Albuminoids.	Purity.
					Per cent.	*Per cent*	*Per cent.*	
September 6..	13	7.70	13.94	1.0566	5.62	5.74	....	47.48
September 7..	19	6.50	11.75	1.0472	5.29	4.26	....	45.01
September 10	31	7.10	12.81	1.0519	6.17	4.55	....	48.16
September 12	37	7.40	13.82	1.0540	7.70	2.33	....	57.80
September 12.	41	6.80	12.20	1.0493	7.44	....	....	60.08
September 13.	46	7.70	13.98	1.0566	8.26	3.13	....	59.71
September 14	53	6.70	12.01	1.0485	6.73	3.06	....	55.89
September 15.	58	7.20	13.00	1.0527	7.07	3.01	....	54.38
September 17	63	7.80	14.00	1.0570	8.30	2.84	.4938	59.28
September 18.	74	6.55	11.82	1.0476	6.74	2.21	.4813	57.10
September 18	73	7.90	14.29	1.0583	8.83	2.39	....	61.79
September 19.	82	6.80	12.25	1.0493	6.70	2.67	.4688	54.79
September 20.	90	8.15	14.75	1.0600	7.54	3.28	.5488	51.10
September 21.	95	7.40	13.41	1.0544	7.39	3.09	.4813	55.10
September 21.	101	7.40	13.28	1.0540	7.41	3.05	....	55.79
September 22	105	6.60	11.91	1.0481	6.06	3.02	....	50.89
September 22.	109	6.90	12.41	1.0502	6.94	2.51	....	56.72
September 24.	115	6.90	12.46	1.0502	7.88	2.06	....	63.24
September 24.	119	7.40	13.28	1.0540	8.40	1.82	.5313	63.25
September 26.	126	7.65	13.85	1.0561	7.69	2.14	....	55.52
September 27.	132	7.50	13.64	1.0553	8.64	2.26	.5625	63.33
September 27.	136	7.40	13.27	1.0540	8.68	1.97	.5125	65.41
September 28.	140	7.40	13.38	1.0544	8.10	2.10	.5000	60.54
September 29	146	7.80	14.01	1.0570	9.16	1.80	.5813	65.24
September 29.	149	8.30	14.94	1.0609	10.16	1.75	....	65.32
September 30	156	7.90	14.24	1.0578	9.08	1.94	....	63.76
September 30.	160	7.80	14.14	1.0576	9.26	1.98	....	65.48
October 2	167	7.50	13.62	1.0553	8.44	2.08	.5560	61.96
October 2	170	7.20	13.03	1.0527	8.71	1.93	....	66.84
October 2	174	7.00	14.18	1.0578	9.35	1.89	....	66.08
October 5	181	7.90	14.15	1.0576	8.13	2.35	....	57.45
October 5	185	8.30	15.00	1.0613	8.50	3.52	.5125	57.26
October 6	193	8.50	15.32	1.0621	9.81	2.24	....	64.03
October 6	199	7.65	13.85	1.0561	8.92	1.83	....	60.07
October 8	208	7.70	13.88	1.0561	8.24	2.03	....	59.30
October 8	213	7.85	13.79	1.0561	8.48	1.96	.5625	61.49
October 9	218	7.80	14.01	1.0570	8.24	2.76	....	58.81
October 11 ...	233	8.30	14.91	1.0609	9.08	2.57	....	60.89
October 11 ...	236	8.30	14.88	1.0608	9.38	2.38	.5750	63.03
October 12 ...	241	6.50	11.74	1.0474	7.11	2.05	....	68.22
October 13	245	8.20	14.82	1.0604	7.17	2.22	....	48.38
October 15	256	8.00	14.44	1.0587	9.64	1.56	.5813	66.76
October 17 ...	265	8.40	15.07	1.0613	8.83	1.84	....	58.59
October 18	270	7.40	13.34	1.0540	8.25	1.52	....	61.84
October 18	274	8.00	14.41	1.0587	8.81	1.93	....	61.13
October 19	279	8.15	14.72	1.0600	8.57	2.17	....	58.22
October 20 ...	285	7.80	14.08	1.0574	9.21	1.64	....	65.41
October 22	289	7.00	12.63	1.0510	8.33	1.17	....	65.95
October 23 ...	294	6.30	11.33	1.0455	7.27	1.28	....	63.87
October 23 ...	298	6.50	11.72	1.0472	7.46	1.22	....	63.65
October 24	307	6.60	11.89	1.0481	7.60	1.31	....	63.81
October 24	310	7.30	13.18	1.0536	8.55	1.24	....	64.87
October 25 ...	316	7.20	12.95	1.0523	8.61	1.16	....	66.44
October 26 ...	320	6.90	12.38	1.0502	8.38	1.16	.5000	67.69
October 27 ...	325	7.50	13.60	1.0553	9.03	1.25	.5625	66.39
October 27 ...	328	7.05	12.70	1.0514	8.52	1.22	....	67.08
October 29	337	7.00	12.63	1.0510	8.20	1.24	....	60.17
October 30	343	6.40	11.60	1.0468	7.40	1.08	....	63.79
October 30	351	7.10	12.79	1.0519	8.42	1.38	....	65.84
November 1...	355	7.65	13.81	1.0561	8.68	1.36	....	62.83
November 2...	362	7.70	12.89	1.0523	7.43	1.74	.5000	57.64
November 2 ..	365	6.70	12.08	1.0489	7.00	.99	....	57.94
November 3...	372	7.00	12.65	1.0510	7.44	1.27	....	58.81
Means	....	7.42	13.37	1.0542	8.07	2.15	.5278	60.41
Maxima...	....	8.40	15.32	1.0621	10.16	5.74	.5813	67.69
Minima...	....	6.30	11.33	1.0455	5.29	.99	.4688	45.01

TABLE V.—*Semi-sirups.*

Date.	No.	Baumé.	Total solids, by Brix, at 15.5°.	Specific gravity.	Sucrose.	Glucose.	Purity.
					Per cent.	*Per cent.*	
Sept. 3......	1	21.50	39.27	1.1789	19.20	....	48 89
Sept. 6......	14	18.70	34.08	1.1521	14.47	11.65	42.45
Sept. 7	20	23.20	42.44	1.1901	18.33	15.75	43.19
Sept. 11......	32	22.30	40.72	1.1835	18.15	12.45	44.57
Sept. 13......	42	23.10	42.22	1.1940	26.61	7.05	63.02
Sept. 14	47	23.00	42.10	1.1940	25.32	8.74	60.14
Sept. 15	55	14.50	26.31	1.1130	13.52	6.39	51.37
Sept. 18	65	24.20	44.21	1.2056	26.60	0.05	60.16
Sept. 20	83	26.40	48.52	1.2294	29.36	9.31	60.51
Sept. 21......	91	28.70	52.94	1.2563	27.14	12.06	51.26
Sept. 23......	102	26.60	48.93	1.2337	28.04	8.85	57.30
Sept. 25......	120	24.06	44.98	1.2003	25.39	9.16	56.44
Sept. 27......	127	27.40	50.41	1.2414	37.41	0.77	74.21
Sept. 28......	137	26.00	47.82	1.2256	29.41	6.33	61.50
Sept. 29......	142	25.80	47.43	1.2240	28.51	7.75	61.09
Sept. 30......	152	25.80	47.41	1.2234	30.57	5.68	64.48
Oct. 2......	164	25.60	46.87	1.2207	31.06	6.46	64.56
Oct. 5......	177	24.80	45.50	1.2077	28.38	5.36	62.37
Oct. 6......	188	23.70	43.40	1.2008	24.68	5.80	56.08
Oct. 7......	200	21.90	40.00	1.1820	23.78	4.81	59.45
Oct. 8......	210	25.45	46.76	1.2196	27.08	6.01	57.99
Oct. 10......	224	25.35	46.55	1.2185	28.90	6.56	62.08
Oct. 12......	238	25.50	47.30	1.2229	30.75	5.35	65.00
Oct. 13......	243	26.20	48.10	1.2272	32.13	7.81	66.79
Oct. 15......	252	26.10	48.00	1.2267	33.98	5.16	70.88
Oct. 16......	257	25.80	47.25	1.2229	34.08	5.41	72.52
Oct. 18......	267	23.50	43.01	1.1987	25.02	5.12	58.17
Oct. 23......	291	26.40	48.53	1.2245	30 97	5.33	63.81
Oct. 24......	300	25.70	47.13	1.2169	31.67	4.89	67.19
Oct. 25......	312	26.10	47.99	1.2212	33.27	4.70	69.32
Oct. 27......	321	23.10	42.34	1.1913	28.26	6.19	66.74
Oct. 27......	330	26.90	49.50	1.2361	33.37	4.12	67.41
Oct. 30......	339	26.70	49.04	1.2272	30.39	5.29	61.97
Nov. 2......	357	23.70	43.25	1.1961	27.18	5.16	62.84
Nov. 5......	376	24.73	44.53	1.2072	30.03	4.55	68.70
Maxima..	......	28.70	52.94	1.2563	37.41	15.75	72.52
Means ...	......	24.55	45.00	1.2092	27.53	7.21	60.70
Minima..	......	14.50	26.31	1.1130	13.52	4.12	42.45

TABLE VI.—*Masse cuite.*

Date.	Number.	Moisture.	Ash.	Glucose.	Sucrose, direct.	Sucrose, indirect.	Solids, not sugar.
		Per cent.	*Per cent.*	*Per cent.*	*Per cent.*	*Per cent.*	
Sept. 15	54	15.62	6.32	21.13	50.40	50.30	6.63
Sept. 18	66	30.40	5.25	17.06	47.00	41.65	5.64
Sept. 19	76	15.80	6.59	16.50	57.40	54.64	6.47
Sept. 21	96	16.28	6.77	19.39	50.10	50.21	5.99
Sept. 21	97	22.52	6.87	20.00	48.40	42.13	7.58
Sept. 24 ..:....	110	14.52	6.46	21.42	52.00	52.41	5.19
Sept. 26	121	14.59	7.08	20.40	52.00	51.72	6.39
Oct. 6	191	14.52	5.55	12.44	63.20	60.65	6.84
Oct. 11	228	14.65	5.10	10.75	65.80	62.94	6.56
Oct. 14	251	....	4.94	10.15	64.00	63.20	....
Oct. 16	258	15.74	5.92	8.73	63.60	63.72	5.89
Oct. 24	301	14.24	5.66	7.34	65.60	63.58	9.18
Oct. 30	344	15.69	6.30	9.35	63.40	63.52	5.14
Nov. 1	377	15.59	7.22	8.31	61.20	60.21	8.07
Maxima...	....	30.40	7.22	21.42	65.80	63.72	....
Means ...	....	16.94	6.15	14.56	57.44	55.78	....
Minima ...	....	14.24	4.94	7.34	48.40	41.65	....

TABLE VII.—*Raw sugars.*

Dates.	No.	Per cent. sugar by polarization.
Sept. 26..	123	77.40
Oct. 8..	201	89.40
Oct. 8..	204	86.00
Oct. 8..	205	84.00
Oct. 13..	242	75.80
Mean...	......	82.52

TABLE VIII.—*First sugars.*

Dates.	No.	Per cent. sugar by polarization.
Oct. 9..	215	97.40
Oct. 9..	381	97.00
Oct. 8..	202	97.80
Oct. 8..	222	98.20
Oct. 9..	223	95.00
Oct. 16..	260	96.00
Oct. 24..	303	95.00
Oct. 30..	346	97.60
Oct. 30.	347	97.00
Oct. 30	381	97.00
Mean...	......	96.80

TABLE IX.—*Molasses.*

Date.	Number.	Moisture.	Ash.	Glucose.	Sucrose, direct.	Sucrose, indirect.	Solids, not sugar.
		Per cent.	*Per cent.*	*Per cent.*	*Per cent.*	*Per cent.*	
Sept 26	122	26.54	7.63	28.41	34.60	35.02	1.80
Oct. 6	190	34.00	7.83	17.33	40.00	39.40	1.61
Oct. 6	196	31.00	7.41	16.78	43.60	43.17	1.64
Oct. 8	203	32.25	7.10	15.76	44.60	43.40	1.40
Oct. 9	221	29.60	6.09	17.33	43.40	43.04	3.94
Oct. 10	225	30.04	7.40	14.57	41.20	41.96	6.03
Oct. 11	229	23.27	6.97	17.30	46.20	44.82	7.64
Oct. 16	259	....	6.85	14.56	42.02	38.25	....
Oct. 24	345	32.03	....	13.32	44.00	40.79	....
Maxima...	....	34.00	7.63	28.41	46.20	44.82	....
Means	....	29.92	7.11	17.26	42.11	41.17	....
Minima ...	....	23.27	6.09	13.32	34.60	35.62	....

TABLE X.—*Mill juices from exhausted chips.*

Date.	No.	Baumé	Brix.	Specific gravity.	Sucrose.	Glucose.	Purity.
Sept. 4.........	7	1.10	2.00	1.0075	.72	....	36.00
Sept. 6.........	12	1.30	2.44	1.0089	.69	.18	28.28
Sept. 7.........	17	1.40	2.54	1.0101	.51	.63	20.08
Sept. 12.........	40	1.60	2.94	1.0113	.81	.51	27.55
Sept. 13.........	44	1.60	2.93	1.0113	1.30	.53	44.36
Sept. 14.........	51	1.50	2.67	1.0103	.73	.05	27.22
Sept. 17.........	64	1.50	2.71	1.0105	.92	.73	33.95
Sept. 18.........	71	1.10	2.03	4.0077	.79	.27	38.91
Sept. 18	75	.90	1.63	1.0062	.69	.25	42.33
Sept. 19.........	81	1.10	1.99	1.0077	.51	.27	25.63
Sept. 20.........	88	1.50	2.68	1.0105	1.23	.40	45.82
Sept. 21.........	93	1.20	2.13	1.0081	.51	.35	23.94
Sept. 21.........	99	1.20	2.22	1.0081	.77	.32	34.18
Sept. 22.........	106	.80	1.38	1.0050	.36	.21	20.09
Sept. 24.........	112	1.40	2.48	1.0097	1.03	.37	41.13
Sept. 24.........	117	1.30	2.35	1.0089	1.08	.30	45.99
Sept. 27.........	130	2.20	4.04	1.0151	1.99	.43	49.25
Sept. 28.........	141	1.80	3.32	1.0125	.77	.36	23.19
Sept. 29.........	144	1.90	3.47	1.0133	2.13	.44	61.38
Sept. 29.........	150	1.50	2.74	1.0105	1.28	.22	46.71
Sept. 30.........	157	1.60	2.90	1.0113	1.28	.23	44.13
Sept. 30.........	161	1.50	2.76	1.0105	1.83	.27	66.30
Oct. 2.........	168	1.40	2.55	1.0097	1.18	.19	46.27
Oct. 2.........	172	1.40	2.61	1.0101	1.36	.20	52.17
Oct. 5.........	182	1.50	2.69	1.0105	1.44	.34	53.53
Oct. 5.........	186	1.10	2.00	1.0077	.72	.23	36.00
Oct. 6.........	194	1.50	2.68	1.0105	1.28	.29	44.03
Oct. 6.........	198	1.70	3.12	1.0117	1.33	.34	42.62
Oct. 8.........	209	1.90	3.47	1.0133	1.58	.52	45.53
Oct. 8.........	214	1.80	3.26	1.0125	1.64	.35	50.30
Oct. 9.........	219	1.50	2.74	1.0105	1.13	.34	41.24
Oct. 11.........	231	1.90	3.47	1.0133	1.49	.43	42.04
Oct. 11.........	237	1.90	3.36	1.0133	1.54	.35	45.81
Oct. 13.........	247	2.90	5.30	1.0209	2.29	.79	43.21
Oct. 15.........	254	2.90	5.32	1.0299	2.55	.52	47.93
Oct. 17.........	260	2.50	4.14	1.0162	1.69	.61	40.82
Oct. 18.........	271	2.70	4.80	1.0189	2.45	.82	51.04
Oct. 18.........	275	2.60	4.60	1.0181	1.74	.52	37.82
Oct. 19.........	280	1.90	3.40	1.0133	1.44	.31	42.35
Oct. 20.........	286	2.80	5.00	1.0197	2.81	.50	56.20
Oct. 22.........	290	1.30	2.40	1.0095	1.18	.19	49.16
Oct. 23.........	295	1.10	2.00	1.0075	1.03	.25	51.50
Oct. 23.........	299	2.20	3.90	1.0151	2.25	.39	57.57
Oct. 24.........	305	1.30	2.40	1.0094	1.33	.23	55.41
Oct. 24.........	311	1.55	2.80	1.0108	1.28	.24	45.72
Oct. 25.........	314	2.40	4.26	1.0165	2.10	.40	49.34
Oct. 26.........	318	2.60	4.62	1.0181	2.40	.33	51.94
Oct. 27.........	329	1.90	3.50	1.0137	1.33	.21	38.00
Oct. 29.........	332	1.70	3.03	1.0117	1.08	.28	35.64
Oct. 29.........	338	1.90	3.46	1.0129	1.69	.26	48.84
Oct. 30.........	341	2.20	3.90	1.0153	1.48	.25	40.51
Nov. 1.........	356	2.40	4.27	1.0169	1.99	.28	46.60
Nov. 2.........	360	2.90	5.17	1.0205	2.91	.27	56.28
Nov. 2.........	366	2.00	3.64	1.0141	1.74	.21	48.66
Nov. 3.........	373	2.20	4.02	1.0157	1.70	.23	44.52
Maxima ..	......	2.90	5.32	1.0299	2.91	.79	66.30
Means	......	1.75	3.17	1.01221	1.40	.36	43.12
Minima ...	......	.80	1.38	1.0050	.36	.18	20.08

TABLE XI.—*Albuminoids.*

Number.	Fresh chips.	Number.	D.ffusion juice.	Number.	Defecated juices.
	Per cent.		*Per cent.*		*Per cent.*
60.........	.6563	72.........	.4688	63.........	.4938
70	.6183	77.........	.4688	74.........	.4813
138........	.6250	79.........	.5438	82.........	.4688
143........	.7438	94.........	.5313	90.........	.5438
165........	.7563	104........	.4688	95.........	.4813
183........	.8063	118........	.5188	119........	.5313
211........	.8875	135........	.5188	132........	.5625
234........	.7313	139........	.5313	136........	.5125
253........	1.0375	145........	.5813	140........	.5000
301........	.4875	166........	.5625	146........	.5813
317........	.8100	184........	.5813	167........	.5500
322........	.6687	212........	.5813	185........	.5125
359........	.9656	235........	.5813	213........	.5625
		255........	.5938	230........	.5750
		297........	.5281	256........	.5813
		306........	.5094	320........	.5000
		319........	.5500	325........	.5625
		324........	.5906	362........	.5000
		333........	.5000		
		361........	.5187		
Means....	.7857	....	.5264	....	.5278

TABLE XII.—*Comparison of acidity in juices from fresh chips and diffusion juices with use of caustic lime.*

Date.	No.	Mill juices from fresh chips. 100 c.c. req. N/10 NaOH.	Sucrose.	Glucose.	Glucose to 100 pts. sucrose.	No.	Diffusion juices. 100 c.c. req. N/10 NaOH.	Sucrose.	Glucose.	Glucose to 100 pts. sucrose.	Extraction.
		c. c.	Pr.cent.	Pr.cent.			c. c.	Pr.cent	Pr.cent.		
Oct. 5	179	32	12.47	3.02	24.2	180	15	7.75	2.60	33.4	88.45
Oct. 5	183	12	10.97	3.66	33.4	184	9	8.09	2.63	32.5	93.40
Oct. 6	189	28	13.97	2.49	17.8	192	12	9.43	2.19	22.1	90.80
Oct. 8	206	38	12.56	2.85	22.7	207	21	8.28	2.08	25.1	87.40
Oct. 8	211	24	12.03	2.31	19.2	212	14	8.50	1.80	22	86.30
Oct. 9	216	39.5	11.97	2.94	24.5	217	33	7.65	2.19	28.6	90.50
Oct. 11	230	35.5	13.28	2.87	21.6	232	18	8.14	2.50	31.8	88.70
Oct. 11	234	30	12.45	2.44	19.6	235	20	9.11	2.22	24.3	87.70
Oct. 13	244	35	10.03	3.91	38.9	246	28	9.48	2.36	24.9	77.20
Oct. 15	253	(*)	15.58	1.31	8.4	255	(*)	9.37	1.55	16.5	83.70
Oct. 19	277	52	14.31	2.61	17.1	278	23.5	8.53	2.13	24.9	......
Oct. 20	283	44.5	13.37	2.19	16.3	284	24.4	8.98	1.54	17.1	......
Oct. 22	287	42	12.47	1.21	9.7	288	14.5	7.93	1.19	15.0	90 50
Oct. 23	292	30	11.08	1.48	12.3	293	15	6.93	1.27	18.3	91.40
Oct. 23	296	25	11.90	1.48	12.3	297	18	7.35	1.25	17.0	81.10
Oct. 24	304	30	12.96	1.29	9.9	306	10	7.20	1.27	17.6	89.70
Oct. 24	308	25	12.19	1.39	11.4	309	10	8.13	1.24	15.2	80.50
Oct. 25	313	38	14.41	1.74	12.1	315	8	8.64	1.19	13.7	85.40
Oct. 26	317	......	14.98	1.00	6.7	319	1	8.58	1.15	13.4	84.00
Oct. 27	322	36	14.29	1.34	9.4	324	15	8.90	1.28	14.3	90.70
Oct. 27	326	19.5	12.91	1.20	9.3	327	7	8.42	1.29	15.3	......
Oct. 29	331	26	14.49	1.18	8.1	333	12	7.42	1.15	15.5	92.50
Oct. 29	335	33	14.01	1.14	8.1	336	13	8.26	1.33	16.1	87.90
Oct. 30	340	26	13.90	1.42	10.1	342	0	7.33	1.20	16.4	89.42
Oct. 30	349	18	13.59	2.10	14.7	350	16	7.20	1.36	18.8	......
Nov. 1	353	20	14.30	2.23	10.6	354	9	8.13	1.40	17.2	86.00
Nov. 2	359	12	13.67	1.99	14.5	361	(†)	7.48	1.80	21.6	77.20
Nov. 2	363	7	12.91	1.13	8.7	364	2	6.92	0.90	14.3	86.50
Means	......	20.2	13.15	1.99	15.4	......	14.4	8.15	1.65	20.2	87.33

TABLE XIII.—*Comparison of acidity of juices without caustic lime*

Sept. 21	92	31.00	9.34	4.12	44.1	94	30.40	6.92	3.12	45.1	94.50
Sept. 22	107	40.80	11.49	3.11	27.1	108	38.50	6.65	2.30	34.6	......
Sept. 24	111	36.00	9.05	3.02	31.3	113	28.20	7.69	2.20	28.6	89.30
Sept. 24	110	40.00	10.89	2.24	20.6	118	28.50	8.20	1.91	23.3	90.00
Sept. 26	124	47.00	12.89	2.51	19.4	125	26.00	6.69	2.23	33.3	......
Sept. 29	143	40.00	12.79	2.45	19.1	145	40.00	8.85	1.96	22.1	83.30
Sept. 29	147	34.50	14.92	1.10	7.9	148	34.50	10.02	1.68	16.7	88.60
Sept. 30	154	49.00	13.82	1.82	13.2	155	49.00	8.86	1.83	20.6	90.70
Sept. 30	158	42.50	13.66	1.83	13.4	159	40.50	9.06	1.01	21.0	88.80
Oct. 2	155	45.10	12.69	2.60	21.2	166	29.70	8.25	2.16	26.1	90.70
Means	......	40.59	12.21	2.49	21.7	......	35.43	8.12	2.13	27.1	89.49

TABLE XIV.—*Acidity and inversion with calcium carbonate (whiting).*

Sept. 27	131	43.4	12.78	2.60	21.0	135	18.00	8.43	2.24	26.5	......
Sept. 27	138	44	13.51	2.14	15.8	139	15.50	7.92	2.24	28.3	......
Means	......	43.7	13.15	2.42	18.4	......	16.75	8.18	2.24	27.4	......

* Neutral.   † Alkaline.

TABLE XV.—*Comparison of Brix spindles with solids by drying.*

MILL JUICES FROM FRESH CHIPS.

Date.	No.	Brix at 15.5°.	Total solids direct film.	Total solids direct with asbestus.	Sucrose.	Glucose.	Purity.	Corrected sucrose.	Corrected purity.
		Pr.cent.	Pr.cent.	Pr.cent.	Pr.cent.	Pr.cent.		Pr.cent.	
Oct. 3	176	18.49	17.21	17.14	12.82	1.61	69.33	12.82	73.62
Oct. 4	183	19.83	18.57	18.15	10.97	3.66	55.32	11.06	60.88
Oct. 5	189	20.40	19.60	19.43	13.97	2.49	68.48	14.03	72.19
Oct. 11	230	21.60	20.10	20.10	13.28	2.87	61.48	13.36	66.49
Oct. 13	244	21.20	19.81	19.73	11.63	3.91	54.86	11.70	59.30
Oct. 18	276	21.81	20.87	20.41	14.84	2.91	68.04	14.92	73.10
Oct. 23	292	17.73	16.60	16.01	11.98	1.48	67.57	12.07	75.40
Oct. 24	304	19.07	17.77	17.25	12.06	1.29	67.06	13.06	75.69
Oct. 25	313	20.34	18.82	18.75	14.41	1.74	70.84	14.50	76.82
Oct. 26	317	21.23	19.56	19.47	14.98	1.00	70.56	15.08	75.80
Oct. 27	322	20.56	18.85	18.86	14.29	1.34	69.50	14.40	76.40
Oct. 29	331	20.60	18.85	18.74	14.49	1.18	73.39	14.60	77.91
Oct. 30	340	20.57	19.06	18.68	13.99	1.42	68.01	14.12	75.57
Nov. 2	359	20.64	18.71	18.13	13.67	1.99	66.23	13.81	76.02
Nov. 3	370	18.67	16.99	17.01	12.86	1.41	68.88	12.95	76.18
Means..	....	20.18	18.72	18.52	13.41	2.02	66.70	13.50	72.76

TABLE XVI.—DIFFUSION JUICES.

Date.	No.	Brix at 15.5°.	Total solids direct film.	Total solids direct with asbestus.	Sucrose.	Glucose.	Purity.	Corrected sucrose.	Corrected purity.
Oct. 4	180	13.65	12.56	12.10	7.75	2.60	56.79	7.83	64.44
Oct. 6	192	14.82	13.36	13.23	9.43	2.10	63.63	9.49	71.75
Oct. 9	217	13.57	12.29	12.21	7.65	2.19	56.66	7.69	63.02
Oct. 11	232	13.85	13.02	12.96	8.14	2.59	58.90	8.16	62.80
Oct. 13	246	14.00	14.04	13.91	9.48	2.36	63.62	9.62	69.36
Oct. 19	278	14.05	12.74	12.29	8.53	2.13	60.61	8.57	60.69
Oct. 23	293	10.78	9.24	9.28	6.89	1.27	63.91	6.93	74.68
Oct. 24	306	11.41	10.47	9.98	7.20	1.27	63.10	7.24	72.37
Oct. 25	315	12.28	11.01	10.91	8.64	1.19	70.35	8.68	79.60
Oct. 26	319	12.38	11.30	11.20	8.58	1.15	69.30	8.62	77.00
Oct. 30	342	11.22	9.89	9.80	7.33	1.20	65.32	7.37	75.19
Nov. 2	361	12.72	11.12	11.16	7.48	1.80	58.91	7.53	67.44
Nov. 3	371	12.47	10.82	10.85	8.33	1.30	66.80	8.39	77.30
Means ..	......	12.93	11.68	11.53	8.11	1.79	62.92	8.16	71.13

REPORT OF HUBERT EDSON, DOUGLASS, KANS.

I herewith submit my report of the work done at Douglass, Kans., during season of 1888.

I wish to call attention to the valuable aid given me by my associate, J. L. Fuelling. Without his assistance much that has been accomplished would not have been done.

Also, I would mention the hearty co-operation of Mr. Fred Hinze in the sugar-house.

After one or two trial runs, to test the machinery of the house, the regular manufacturing season at Douglass commenced September 14, and continued, with what regularity was possible, up to October 25.

There is no doubt but that the Early Amber was ready for work by the middle of August and possibly earlier. When I arrived in Douglass, August 26, I found several fields that had passed maturity. This cane, however, contrary to experience elsewhere, did not deteriorate in any marked degree till some time after reaching its maximum sucrose. When the house was closed we still had Amber coming in in large quantities, and containing sucrose enough to warrant working it.

Besides the Amber the two other varieties chiefly grown were the Orange, and a cane identified by Mr. Denton, of Sterling, Kans., as the Chinese.

The Amber and Chinese contained highest sucrose and lowest glucose, with the advantage slightly in favor of the Chinese. The Orange did not do as well as was expected, but it was planted so late in the season that it did not have time to mature.

The exceedingly variable nature of the cane brought in was a source of constant annoyance, nor would the appearance of the stalks be any criterion of the quality of the juice. One field of 30 acres which had been ordered hauled in before any test had been made of it was found on the arrival of the first load to contain but 4.50 per cent. sucrose, with almost as much glucose. This cane was, judging by its appearance, as good as any worked during the season, but repeated tests of samples taken from different parts of the field failed to show in a single instance enough sucrose to warrant working for sugar. Numerous instances of this same thing were found throughout the season, and the cane needed the closest watching.

94

One thing it would be well to impress upon the sorghum grower, and that is the necessity of growing small or medium sized canes. From numerous trials of comparative samples the highest sucrose and lowest glucose were always found in the smaller canes. Fields also where the small and slender canes predominated were always of superior quality. The best cane analyzed at Douglass was a sample from a field sowed for fodder, in which the seed had been scattered broadcast on the land, and as a consequence grew very small. Of course I do not mean to advocate the sowing of sorghum seed to grow a product for the sugar-house, as then too large an amount of sheath and leaves would be obtained, but it is necessary to avoid large rank stalks if the desire is to obtain a high content of sucrose.

SUGAR-HOUSE.

The house was designed to work 100 tons of field cane daily. The Hughes cutter and shredder were used. The trap-door just before the cutter, through which it was intended to pass the seed heads, failed to work satisfactorily. This was due, in part at least, to the heavy feed which it was necessary to keep on the narrow carriers in order to supply the battery with chips. The shredder when properly adjusted did excellent work, tearing the chips into a pulp if required.

The main feature of the house was the diffusion battery. This is known as the Hughes system of diffusion, and is described in Bulletin 17, chemical division, Department of Agriculture. The one at Douglass differed slightly, however, from the one described there. The main battery contained ten cells, with the baskets for holding chips used in his process, and in addition to these an outside cell was placed so that the arm from the large crane could reach the basket while immersed in it.

An extra crane was necessary to raise and lower the baskets in this cell, as it had to be worked without connection with the main battery. The object of the cell was to give a dense diffusion juice and thus save evaporation. As the battery progressed the heaviest juice from two cells were drawn into the outside cell, and there received two baskets of fresh chips before being discharged. This, as far as I was able to see, did not attain the object claimed for it, as no fresh chips ever reached the main battery, and consequently the juices were more dilute and needed the addition of two baskets of fresh chips to bring them to a normal diffusion juice. It is certain at least that the extra steam-power required to run the outside cell would a great deal more than suffice to evaporate any less dense juice that might be obtained.

Before passing to the work done by the battery, as a whole it is but just to say that there were mechanical defects in the construction which if they could have been remedied this season would have materially assisted the quality of the work. The bottom of the baskets, instead of being single and swinging to one side, were double and hinged to a cross-bar extending from one side of the basket to the other. As a consequence of this arrangement the emptying of the exhausted chips was a very difficult matter. But, on the other hand, a basket constructed strong enough to permit a single bottom would be altogether too heavy to use where so much of the work is done by hand.

The average sucrose of the fresh chips for the season was 9.88; for the exhausted chips, 1.72. The extraction of sucrose, therefore, was $9.88-1.72=8.16\div9.88=82.59$ per cent. This extraction was accompanied by a dilution of 52.45 per cent. 16.89 (Brix of fresh chips)-8.03 (Brix of diffusion juice); $8.86\div16.89=52.45$ per cent. With a dilution of this sort in a closed battery practically all the sugar would be exhausted instead of 1.72 per cent. left in by the Hughes process.

It was noticed that a regular ratio existed between the exhaustion and the dilution. As the dilution was increased the extraction became better, and *vice versa.*

Besides the amount of sugar left in the chips there was an unknown waste of immense quantities of juice from the drippings of the baskets in transferring them from the eleventh cell to the cells of the main battery. This loss it was impossible to gauge, but to any one who saw it, it was evident that no inconsiderable amount was lost.

Nothing which we could think of to make the battery a success was left undone. For part of the time I shifted all of the laboratory work to my associate, Mr. Fuelling, and took charge of the battery. This I was prepared to do from a previous year's work with the inventor of the system, with whose plan of running the battery I was consequently familiar. Although the quality of the work was improved after the change I instituted, it was so far from being good diffusion, that nothing was left to do but to condemn the apparatus.

THE DIFFUSION JUICE.

The juice as it came from the cells was full of finely-divided fiber which had come through the perforations of the baskets, and was also of such a dirty black color that it was impossible to clarify it.

Sulphites of lime were used for awhile, as were also superphosphates, but both were so full of sulphuric acid and accomplished so little, that they were discontinued.

The juice probably acquired some of this color from its acids attacking the iron vessels in which it was kept so much of the time, but the main cause was the passage of large quantities of seeds through to the diffusion battery along with the fresh chips. As was mentioned before, the cutter was too narrow for the capacity of the house, and a very heavy feed was kept on the carrier, preventing the seed-heads dropping down through the trap-door designed for that purpose.

To illustrate that these seeds were the cause of the discoloration, Mr. Fuelling diffused two beakers full of chips, the one of them containing a few seed and the other none.

The one with the seed gave the black color characteristic of the diffusion juice from the house, while the other gave a perfectly clear limpid liquor. I endeavored to have the superintendent of the house make a run, cutting the tops off in the field, but he failed to do so.

DISPOSITION OF EXHAUSTED CHIPS.

During the first part of the season a long carrier was used to convey the chips to the yard. It was intended to extend this as the yard filled up, but the chains broke so often, that this plan was given up and the chips taken off in carts.

The centrifugals did very poor work throughout the season; but so little sugar was extracted by the battery that it was not considered necessary to get new ones.

SUMMARY OF WORK.

During the season 2,167 tons of cane were worked. Allowing 25 per cent. off for tops and leaves, this would amount to 1,623 tons of cleaned cane.

Forty-five thousand pounds of sugar, 94.45 polarization, were obtained, or 26.2 pounds per ton of clean cane.

Eliminating the loss in the centrifugals, which would have been remedied if enough sugar had been obtained to justify it, the great loss in working the house was in the battery.

RESULTS OF ANALYSES.

TABLE No. XVII.—*Sorghum cane.*

Date.	No.	Total solids by Brix at 17.5°	Specific gravity.	Sucrose.	Purity.	Character of sample.
				Per ct.		
Sept. 5	......	16.34	1.0669	9.47	57.95	Blown-down Amber.
Sept. 5	......	16.43	1.0674	9.32	56.73	Standing upright.
Sept. 6	......	15.84	1.0648	5.34	33.97	Left on carrier for five days.
Sept. 6	......	17.62	1.0727	5.34	30.31	Do.
Sept. 6	......	22.14	1.0882	12.44	56.22	Do.
Sept. 7	......	15.20	1.0621	9.61	63.25	Amber from Holmes's farm.
Sept. 8	......	16.54	1.0675	6.15	37.12	Left on carrier for seven days.
Sept. 8	......	18.21	1.0753	13.47	73.97	Medium-sized Amber from upland.
Sept. 8	......	18.10	1.0748	10.83	59.83	Medium-sized Amber from lowland.
Sept. 8	......	17.64	1.0727	13.16	74.60	Cane from lowland.
Sept. 11	13	18.50	1.0766	12.87	69.52	Sport cane, 11 feet 5 inches long.
Sept. 11	14	17.22	1.0709	3.73	21.66	Cane from carrier.
Sept. 12	19	20.63	1.0855	2.93	14.20	Cut for two days.
Sept. 12	23	17.92	1.0739	12.38	68.97	One sport.
Sept. 12	24	17.93	1.0739	12.88	71.84	Amber.
Sept. 13	28	17.10	1.0704	11.37	66.49	Chinese.
Sept. 13	29	18.37	1.0757	11.65	63.42	White African.
Sept. 13	30	17.67	1.0726	11.86	67.12	Cane from Couch's.
Sept. 13	31	17.98	1.0739	13.30	74.97	Standing from Mr Algiers.
Sept. 13	32	17.47	1.0717	13.18	75.45	Fallen from Mr. Algiers.
Sept. 13	33	17.50	1.0722	9.32	53.25	Amber from carrier.
Sept. 13	34	16.28	1.0662	11.45	70.33	Orange.
Sept. 13	40	18.00	1.0663	10.21	68.07	Do.
Sept. 13	41	18.67	1.0770	12.00	61.26	Amber.
Sept. 13	42	17.69	1.0726	10.76	60.82	Do.
Sept. 13	41	15.83	1.0617	10.57	66.77	Orange.

TABLE No. XVII.—*Sorghum cane*—Continued.

Date.	No.	Total solids by Brix at 17.5°	Specific gravity.	Sucrose.	Purity.	Character of sample.
				Per ct.		
Sept. 14	49	15.27	1.0021	9.09	59.55	Mixed canes.
Sept. 14	50	8.52	1.0327	3.75	44.23	Orange.
Sept. 14	51	15.77	1.0643	10.20	64.64	Do.
Sept. 14	52	16.33	1.0669	11.99	79.54	Do.
Sept. 14	54	16.00	1.0656	10.29	64.31	Amber from carrier.
Sept. 15	63	17.83	1.0739	12.41	69.21	Chinese.
Sept. 17	77	17.54	1.0792	11.48	65.45	Orange.
Sept. 17	81	11.57	1.0464	5.13	42.61	Mixed Amber.
Sept. 17	82	15.97	1.0652	9.81	61.42	Do.
Sept. 17	83	15.74	1.0643	10.88	69.12	Do.
Sept. 17	84	14.11	1.0574	8.50	60.74	Do.
Sept. 17	85	18.32	1.0757	12.28	67.03	Do.
Sept. 17	86	17.87	1.0735	12.03	67.32	Do.
Sept. 17	87	16.27	1.0665	9.55	58.69	Do.
Sept. 17	88	19.05	1.0788	12.87	67.53	Do.
Sept. 17	92	16.72	1.0687	10.10	60.41	Do.
Sept. 17	94	18.21	1.0733	9.65	52.99	Do.
Sept. 18	95	15.70	1.0643	9.48	60.04	Do.
Sept. 18	96	12.40	1.0502	7.15	57.66	Orange.
Sept. 18	105	15.38	1.0626	8.97	58.32	Amber.
Sept. 19	106	14.22	1.0578	8.48	59.63	Jersey Orange.
Sept. 19	107	16.19	1.0660	9.68	59.79	Sprouts from above.
Sept. 19	108	13.60	1.0553	8.06	59.26	Kansas Orange.
Sept. 19	109	18.92	1.0783	12.07	66.06	Mixed Amber.
Sept. 19	110	10.77	1.0430	5.40	50.14	Late Orange.
Sept. 19	112	21.17	1.0909	15.08	71.23	Amber.
Sept. 19	113	19.83	1.0824	14.14	71.30	Chinese.
Sept. 19	114	16.48	1.0695	11.52	67.84	Orange.
Sept. 21	117	9.77	1.0388	4.71	47.18	Do.
Sept. 21	118	10.57	1.0422	5.30	50.14	Do.
Sept. 22	127	13.83	1.0561	9.08	65.73	Mixed Amber.
Sept. 23	141	17.72	1.0730	11.33	63.93	Sport cane.
Sept. 26	177	17.90	1.0739	12.46	69.55	Mixed cane.
Sept. 26	178	17.75	1.0732	12.18	68.50	Do.
Sept. 26	187	15.06	1.0613	8.61	57.17	Cane red at heart.
Sept. 26	189	15.77	1.0643	8.76	55.55	Orange.
Sept. 26	190	17.23	1.0709	11.51	66.80	Amber.
Sept. 28	203	10.92	1.0439	4.75	43.49	Orange.
Sept. 28	204	20.24	1.0841	15.60	77.07	Small Orange, planted close.
Sept. 29	206	19.30	1.0801	14.65	76.00	Mixed cane.
Sept. 29	207	18.49	1.0761	11.72	63.32	Cane from carrier.
Oct. 1	212	19.14	1.0806	13.36	75.02	Amber.
Oct. 1	213	19.19	1.0828	14.37	74.88	Chinese.
Oct. 2	227	18.10	1.0718	13.58	75.02	Amber.
Oct. 2	228	18.25	1.0753	6.36	34.85	Orange.
Oct. 3	245	18.03	1.0744	12.61	69.99	Do.
Oct. 3	246	13.20	1.0536	6.53	49.39	Do.
Oct. 4	257	15.54	1.0634	9.49	61.00	Amber.
Oct. 4	258	14.70	1.0600	6.21	42.00	Do.
Oct. 5	272	17.73	1.0730	11.46	64.74	Do.
Oct. 5	273	17.06	....	11.98	70.22	Do.
Oct. 5	274	20.30	....	15.03	74.04	Small cane.
Oct. 5	275	18.00	....	12.37	68.72	White African.
Oct. 5	276	19.00	....	12.99	68.36	Orange.
Oct. 5	277	16.00	....	10.88	65.54	Do.
Oct. 6	285	13.82	....	8.15	58.97	Do.
Oct. 6	287	18.34	....	13.13	71.59	Do.
Oct. 8	300	13.35	....	7.58	56.78	Do.
Oct. 8	301	16.07	....	11.00	65.98	Do.
Oct. 8	308	15.20	....	9.09	60.10	Do.
Oct. 10	322	12.05	....	5.05	45.94	Amber.
Oct. 12	342	12.19	....	5.51	45.20	Orange.
Oct. 12	343	17.60	....	11.21	63.70	White African.
Oct. 15	371	19.23	....	12.42	64.58	Amber.
Oct. 18	376	9.54	....	4.64	48.67	Orange.
Oct. 19	391	15.10	....	8.06	59.39	White African.
Oct. 19	392	12.75	....	5.55	43.48	Orange.
Oct. 20	398	14.27	....	6.88	48.31	Orange, first frost.
Oct. 20	399	17.77	....	12.13	71.75	Orange.
Oct. 20	400	17.60	....	12.13	69.92	Do.
Oct. 20	401	15.03	....	8.07	51.63	Do.
Oct. 20	402	13.78	....	8.62	61.82	Do.
Oct. 20	403	14.56	....	8.60	52.10	Do.
Oct. 20	404	15.69	....	9.54	60.80	Do.
Oct. 20	405	17.62	....	11.85	67.25	Do.
Oct. 20	406	16.38	....	9.75	61.96	Do.
Mean		16.39	1.0681	10.05	60.64	

TABLE XVIII.—*Fresh chips.*

Date.	No.	Total solids by Brix at 17.5°.	Sucrose.	Purity.	Glucose.	Albuminoids.
			Per cent.	*Per cent.*		*Per cent.*
Sept. 14	56	16.50	0.51	57.64	3.25	....
Sept. 14	60	18.71	9.75	52.00	4.59	....
Sept. 15	61	16.25	8.85	54.86	3.86	....
Sept. 15	69	18.00	9.20	51.11	3.70	....
Sept. 16	73	17.77	10.67	60.04	3.03	....
Sept. 16	78	18.86	12.33	65.37	2.93	....
Sept. 17	90	16.51	9.73	58.93	3.24	....
Sept. 17	97	21.34	15.02	70.38	2.37	....
Sept. 17	101	19.25	12.90	67.01	2.16	....
Sept. 21	119	13.32	5.40	40.54	2.72	....
Sept. 22	123	15.75	9.10	57.78	2.90	....
Sept. 22	130	15.20	9.61	63.22	2.81	....
Sept. 23	134	15.17	7.86	51.81	3.21	....
Sept. 23	138	17.13	11.12	61.91	1.40	....
Sept. 23	143	17.11	10.88	63.59	2.81	....
Sept. 23	146	17.79	10.50	59.36	2.58	....
Sept. 24	150	16.61	10.45	62.91	2.58	....
Sept. 24	154	16.10	9.10	56.52	3.07	....
Sept. 24	158	16.82	10.55	62.72	2.60	....
Sept. 25	165	15.23	7.85	51.54	3.93	....
Sept. 25	169	17.03	10.46	61.41	2.95	....
Sept. 25	173	18.25	11.25	67.12	2.27	....
Sept. 26	179	17.49	10.69	61.12	2.91	....
Sept. 26	183	17.20	9.75	56.76	3.12	....
Sept. 27	191	15.37	7.72	50.26	3.47	....
Sept. 27	195	19.25	11.79	60.21	3.85	....
Sept. 27	199	17.21	9.72	56.48	3.48	....
Oct. 1	208	18.22	11.20	61.96	2.37	....
Oct. 1	214	16.60	8.87	53.43	3.55	....
Oct. 2	217	18.18	11.74	64.57	2.33	....
Oct. 2	220	16.50	8.56	51.08	3.80	.51875
Oct. 2	223	16.10	10.47	65.03	2.14	....
Oct. 2	229	14.43	8.56	59.37	2.55	....
Oct. 3	232	14.72	9.37	64.05	2.45	....
Oct. 3	235	15.69	9.51	60.61	2.59	.57187
Oct. 3	242	16.45	9.63	58.54	3.15	....
Oct. 3	247	16.60	9.19	55.36	3.22	....
Oct. 4	252	17.78	10.03	56.51	3.23	....
Oct. 4	259	18.43	11.20	60.77	2.71	....
Oct. 4	263	17.32	9.88	57.01	2.95	....
Oct. 5	267	17.48	8.90	51.43	4.20	....
Oct. 6	278	16.07	9.19	57.06	3.03	.53125
Oct. 6	288	16.81	9.12	54.25	3.13	....
Oct. 7	292	17.53	11.81	67.37	2.21	....
Oct. 7	200	17.01	10.88	64.55	2.35	....
Oct. 8	304	16.88	9.76	57.88	2.85	.78750
Oct. 9	309	15.41	8.20	53.83	3.30	....
Oct. 9	314	19.41	11.30	58.22	3.35	1.04378
Oct. 10	318	16.79	9.83	58.55	3.00	.62813
Oct. 10	327	16.81	9.37	55.74	3.16	....
Oct. 11	332	17.00	9.32	51.82	3.45	.60312
Oct. 11	337	18.58	12.35	66.47	2.25	....
Oct. 12	344	18.11	10.95	60.46	2.95	.62812
Oct. 12	350	17.70	10.88	61.47	2.73	....
Oct. 15	355	15.31	8.81	56.90	2.95	....
Oct. 15	359	18.70	11.64	61.71	2.65	.70312
Oct. 15	363	15.46	8.24	53.30	3.80	....
Oct. 17	367	17.99	10.60	50.17	3.06	....
Oct. 17	372	16.06	8.97	55.85	3.00	....
Oct. 18	377	17.47	10.48	60.43	3.88	....
Oct. 19	387	17.87	9.57	51.16	3.43	.63125
Oct. 20	394	14.18	7.26	51.24	3.62	.35750
Oct. 20	407	15.50	8.33	53.77	2.98	....
Oct. 21	414	13.56	7.32	54.01	2.81	.46250
Oct. 23	424	18.15	10.98	60.50	3.09	.51325
Oct. 23	428	16.05	8.84	55.34	2.93	....
Oct. 24	432	15.23	8.63	56.70	2.81	.59062
	Mean.	16.89	9.88	58.34	3.01	.61648

100

TABLE XIX.—*Diffusion Juice.*

Date.	No.	Total solids by Brix at 17.5°.	Sucrose.	Purity.	Glucose.	Albuminoids.
			Per cent.		*Per cent.*	*Per cent.*
Sept. 14	57	7.05	4.16	59.01	1.60	....
Sept. 15	61	6.37	3.92	61.54	1.21	....
Sept. 15	65	5.13	2.71	52.77	1.26	....
Sept. 16	70	4.38	2.67	60.95	.83	....
Sept. 16	74	5.44	3.40	62.93	.98	....
Sept. 17	79	5.52	3.57	64.67	1.00	....
Sept. 17	91	6.47	4.07	62.90	1.11	....
Sept. 18	98	6.04	4.21	69.62	.83	....
Sept. 18	102	6.00	3.95	65.86	.78	....
Sept. 21	120	4.81	2.78	57.79	.94	....
Sept. 22	124	6.55	3.94	60.15	1.19	....
Sept. 22	131	4.57	2.77	60.61	.94	....
Sept. 23	135	5.45	3.19	54.53	1.06	....
Sept. 23	139	6.31	4.13	68.33	1.00	....
Sept. 23	144	5.40	3.53	65.37	.82	....
Sept. 23	147	5.86	3.72	63.48	1.03	....
Sept. 24	151	7.04	4.49	63.78	1.08	....
Sept. 24	155	6.48	3.94	60.49	1.15	....
Sept. 24	159	6.00	3.62	60.33	1.05	....
Sept. 25	162	5.33	3.72	47.72	1.18	....
Sept. 25	166	6.02	4.12	59.53	1.10	....
Sept. 25	170	6.86	4.04	58.89	1.20	....
Sept. 25	174	8.94	5.24	58.60	1.28	....
Sept. 26	180	9.00	5.31	59.10	1.60	....
Sept. 26	184	8.45	4.92	58.23	1.49	....
Sept. 27	192	10.25	5.91	58.14	1.51	....
Sept. 27	196	11.63	6.45	57.18	2.26	....
Sept. 27	200	9.49	5.43	57.22	1.87	....
Oct. 1	209	9.00	5.80	65.11	1.53	....
Oct. 1	215	8.49	4.72	55.50	1.61	....
Oct. 2	218	8.01	4.74	59.17	2.57	....
Oct. 2	221	8.30	4.46	53.73	1.53	.26250
Oct. 2	224	8.21	4.90	59.68	1.25	....
Oct. 2	230	7.10	4.06	57.18	1.30	....
Oct. 3	236	7.66	4.28	55.87	1.38	.22500
Oct. 3	243	6.55	3.41	52.06	1.86	....
Oct. 3	248	7.91	4.40	55.62	1.31	....
Oct. 4	253	9.12	5.00	55.81	1.44	....
Oct. 4	260	9.06	4.99	55.07	1.42	....
Oct. 4	264	9.55	5.15	53.92	......	....
Oct. 4	268	9.75	5.61	57.54	1.42	....
Oct. 6	279	9.89	5.13	51.78	1.83	.33437
Oct. 6	289	10.79	6.30	58.38	1.80	....
Oct. 7	293	10.07	5.92	58.78	1.68	....
Oct. 7	297	11.07	7.02	63.41	1.62	....
Oct. 8	305	11.21	6.67	59.50	1.74	.33750
Oct. 9	310	10.83	6.15	56.79	1.91	....
Oct. 9	315	9.82	5.45	55.49	1.88	.33450
Oct. 10	319	9.95	5.87	58.99	1.44	.33125
Oct. 10	328	9.14	5.18	56.67	1.76	....
Oct. 11	333	10.15	5.18	51.03	2.02	.35000
Oct. 11	338	8.79	5.60	63.70	1.30	....
Oct. 12	345	7.58	4.45	58.71	1.18	.23125
Oct. 12	351	7.80	4.91	64.23	1.24	....
Oct. 13	356	9.00	5.62	62.67	1.30	....
Oct. 15	360	10.54	6.11	57.97	1.78	.38775
Oct. 15	364	9.07	5.26	57.97	1.62	....
Oct. 17	368	7.52	4.51	62.07	1.14	....
Oct. 17	376	9.34	5.31	57.82	1.59	....
Oct. 18	378	9.38	5.21	55.55	2.10	.34687
Oct. 19	385	7.70	4.30	54.31	1.51	.28137
Oct. 20	395	8.39	4.28	51.13	2.08	.28125
Oct. 20	408	8.25	4.94	59.91	1.57	....
Oct. 21	415	8.23	4.66	56.72	1.56	.25625
Oct. 23	425	9.22	5.43	58.91	1.75	.27187
Oct. 23	429	7.75	3.99	51.61	1.76	....
Oct. 24	433	8.34	4.80	57.55	1.72	.31250
	Mean.	8.00	4.69	58.63	1.44	.30248

TABLE XX.—*Clarified juice.*

Date.	No.	Total solids by Brix at 17.5°.	Sucrose.	Purity.	Glucose.	Albuminoids.	
			Per cent.		*Per cent.*	*Per cent.*	
Oct. 4....	254	9.32	5.26	57.52	1.48	....	
Oct. 4....	261	9.36	4.98	51.27	1.68	....	
Oct. 4....	265	9 97	5.42	54.36	1.69	....	
Oct. 5....	269	10.10	5.71	56.54	1.46	....	
Oct. 6....	280	14.45	7.68	53.15	2.60	.43437	
Oct. 6....	290	12.33	6.76	54.83	2.20	....	
Oct. 7....	294	12.98	7.59	58.63	2.24	....	
Oct. 7....	298	11.56	7.23	64.29	1.48	....	Lime.
Oct. 8....	306	11.04	6.35	57.22	1.80	.30625	Sulphite used.
Oct. 9....	311	11.86	6.57	56.24	1.91	....	
Oct. 9....	316	10.34	5.74	55.51	1.72	.32360	
Oct. 10....	320	10.61	6.01	56.64	1.68	....	
Oct. 10....	329	10.00	5.42	54.20	1.76	.324'0	
Oct. 11....	334	10.55	6.37	60.90	2.16	.33750	
Oct. 11....	330	9.05	5.57	61.37	1.30	.25625	
Oct. 12....	346	7.02	4.61	58.21	1.20	....	
Oct. 12....	352	8.04	4.75	59.08	1.14	....	
Oct. 13....	357	9.22	5.59	60.63	1.44	....	
Oct. 15....	361	11.23	6.34	56.46	1.86	.42500	
Oct. 15....	365	10.41	5.46	52.54	1.84	....	
Oct. 17	369	8.14	4.88	59.99	1.09	....	
Oct. 17....	374	9.45	5.26	55.06	1.60	....	
Oct. 18....	379	10.08	5.52	44.81	2.02	.31250	
Oct. 19....	389	7.64	4.17	54.77	1.43	.28125	
Oct. 20....	396	8.90	4.63	51.72	2.04	.27187	
Oct. 20....	409	8.47	4.96	58.67	1.55	....	No lime used.
Oct. 21....	416	8.32	4.57	51.99	1.60	.26502	
Oct. 23....	420	9.24	5.57	60.36	1.82	.27185	
Oct. 23....	430	7.90	3.97	50.34	1.75	....	
Oct. 24....	434	8.77	4.49	51.20	1.77	.33125	
Highest		14.45	7.68	64.29	2.60	.43437	
Average		9.91	5.55	55.84	1.71	.31871	
Lowest		7.64	3.97	44.81	1.00	.25625	

TABLE XXI.—*Semi-sirup.*

Date.	No.	Total solids by Brix at 17.5°.	Sucrose.	Purity.	Glucose.	Date.	No.	Total solids by brix at 17.5°.	Sucrose.	Purity.	Glucose.
			Per ct.		*Per ct.*				*Per ct.*		*Per ct.*
Sept. 15.	59	41.10	22.17	53.17	9.09	Sept. 26.	186	84.00	48.52	57.76	16.01
Sept. 15.	67	34.10	18.32	53.43	7.69	Sept. 27.	198	37.15	22.31	60.05	6.41
Sept. 16.	72	34.87	10.17	54.97	6.99	Sept. 27.	204	39.68	23.03	58.04	7.54
Sept. 16.	76	......	17.03	......	7.71	Oct. 1.	211	35.86	21.56	00.12	6.82
Sept. 18	94	60.46	26.77	53.05	8.04	Oct. 2.	226	37.68	21.90	58.13	7.04
Sept. 18.	100	37.38	23.76	63.56	5.61	Oct 3	238	39.58	23.71	59.90	6.73
Sept. 18.	104	38 42	24.80	64.55	5.35	Oct. 4.	256	38.05	22.02	57.92	6.50
Sept. 22.	122	35.68	17.34	48.59	9.86	Oct. 5.	271	38.47	22.73	59.08	6.92
Sept. 22.	126	30.12	20.12	51.43	9.55	Oct. 9.	313	36.92	23.08	62.51	6.13
Sept. 23.	133	33.10	13.05	57.45	6.65	Oct. 10.	331	43.03	25.62	58.32	8.23
Sept. 23.	137	36.03	20.85	57.87	6.90	Oct. 11.	341	43.95	25.68	58.43	8.26
Sept. 23	142	31.90	19.71	61.79	5.96	Oct. 13.	354	39.50	23.42	59.29	6.79
Sept. 24.	149	37.08	23.49	61.85	6.44	Oct. 19.	393	......	23 33	......	8.81
Sept. 24.	153	40.30	25.59	63.50	7.30	Oct. 21.	418	48.10	28.40	59.05	9.39
Sept. 24.	157	36.72	22.73	61.01	6.87						
Sept. 25.	164	37.04	20.53	55.42	6.72	Highest	.....	81.00	48.52	64.55	16.01
Sept. 25.	172	56.46	33.54	59.40	12.49	Average	.....	41.22	23.79	57.99	7.81
Sept. 26.	176	45.60	25.16	55.18	7.85	Lowest .	.....	31.90	17.34	48.59	5.35
Sept. 26.	182	48.57	29.02	51.03	9.11						

TABLE XXII.—*Masse cuite.*

Date.	No.	Sucrose.	Sucrose, double pol.	Glucose.	Moisture.	
		Per cent.	*Per cent.*	*Per cent.*	*Per cent.*	
Sept. 12	22	39.00	....	21.10	....	First strike.
Sept. 22	128	53.44	....	15.43	17.59	Second strike.
Oct. 3	2:9	54.34	....	17.05	16.50	Fourth strike.
Oct. 8	282	51.31	....	15.29	20.85	Fifth strike.
Oct. 8	285	55.69	57.20	15.57	20.79	Seventh strike.
Oct. 8	303	51.60	55.00	18.93	16.86	Eighth strike.
Oct. 10	323	52.60	55.00	16.08	19.24	Sixth strike.
Oct. 10	326	53.50	....	16.77	19.10	Ninth strike.
Oct. 18	381	50.00	52.80	16.43	22.25	Tenth strike.
Oct. 18	384	49.00	52.80	16.60	24.70	Eleventh strike.
Oct. 20	411	50.20	55.00	17.69	21.43	Twelfth strike.
Oct. 21	419	51.60	52.80	18.35	17.69	Thirteenth strike.
Oct. 30	439	44.00	46.64	19.02	23.00	Second masse cuite.
Highest		55.60	57.20	21.10		
Average		50.48	53.41	17.25	20.00	
Lowest		39.00	46.64	15.29	....	

TABLE XXIII.—*Sugar.*

Sept. 19	111	89.80	....	....	....	First strike.
Sept. 29	129	93.51	....	....	....	Second strike.
Oct. 3	240	95.04	....	....	....	Fourth strike.
Oct. 6	283	95.00	....	....	....	Fifth strike.
Oct. 10	324	91.08	....	....	....	Sixth strike.
Oct. 12	348	95.00	....	....	....	Seventh strike.
Oct. 18	383	93.60	....	....	....	Tenth strike.
Oct. 18	385	93.60	....	....	....	Eleventh strike.
Oct. 20	413	96.00	....	....	....	Twelfth strike.
Oct. 21	421	96.40	....	....	....	Thirteenth strike.
Oct. 21	423	94.40	....	....	....	
Oct. 21	437	99.60	....	....	....	Reboiled sugar.
Oct. 21	440	94.00	....	....	....	Second sugars.
Highest		99.60	....	....	....	
Average		94.45	....	....	....	
Lowest		89.80	....	....	....	

TABLE XXIV.—*Molasses.*

Oct. 3	241	36.11	....	17.27	22.86	Fourth strike.
Oct. 6	284	38.60	....	19.76	28.85	Fifth strike.
Oct. 10	325	38.40	41.80	18.11	27.66	Sixth strike.
Oct. 12	349	37.60	39.60	21.29	23.50	Seventh strike.
Oct. 18	382	44.00	....	22.48	22.84	Tenth strike.
Oct. 18	385	40.00	48.40	....	23.05	Eleventh strike.
Oct. 20	412	39.10	41.80	17.77	27.76	Twelfth strike.
Oct. 21	420	38.40	41.14	18.40	26.05	Thirteenth strike.
Oct. 21	422	42.00	44.00	20.05	23.29	Ninth strike.
Oct. 30	441	42.00	41.00	17.40	27.08	From seconds.
Highest		46.00	48.40	22.48	....	
Average		40.16	42.96	19.17	25.30	
Lowest		36.11	39.60	17.27	....	

TABLE XXV.—*Ex chips.*

Date.	No.	Total solids by Brix at 17.5°.	Sucrose.	Purity.	Glucose.	Date.	No.	Total solids by Brix at 17.5°.	Sucrose.	Purity.	Glucose.
			P. ct.		*P. ct.*				*P. ct.*		*P. ct.*
Sept. 14.	58	3.81	1.43	37.53	.56	Oct. 3..	249	3.36	1.50	47.35	.56
Sept. 15.	62	3.04	1.43	39.20	.79	Oct. 4..	255	4.72	2.35	49.81	.75
Sept. 15	68	2.00	1.29	44.48	.50	Oct. 4..	262	4.70	2.38	50.64	.72
Sep. 16.	71	2.34	.07	41.45	.35	Oct. 4..	266	3.56	1.62	45.51	.61
Sept. 16.	75	3.10	1.42	45.81	.45	Oct. 5 .	270	3.08	1.64	45.65	.53
Sept. 17	80	3.73	1.60	45.31	.55	Oct. 6..	281	6.00	2.84	46.07	.90
Sept. 18	99	3.88	2.05	52.81	.34	Oct. 6.	291	6.17	3.58	58.02	1.06
Sept. 21.	121	2.51	.87	34.80	.81	Oct. 7..	205	5.64	3.06	51.25	.63
Sept. 22	125	2.75	1.59	57.82	.44	Oct. 7.	290	5.76	2.70	46.87	.62
Sept. 22	132	1.05	.69	45.63	.33	Oct. 8..	307	4.53	2.25	49.07	.68
Sept. 23.	136	2.34	1.04	44.44	.40	Oct. 9..	312	4.58	2.39	52.18	.69
Sept. 23.	140	3.37	1.58	46.88	.53	Oct. 9..	317	4.23	2.03	47.99	.70
Sept. 23	148	3.25	1.52	46.77	.49	Oct. 10	321	4.29	2.03	47.32	.69
Sept. 24	152	2.76	1.44	52.17	.36	Oct. 10.	330	3.03	1.93	40.11	.66
Sept. 24	156	2.70	1.26	46.66	.47	Oct. 11..	335	5.12	2.50	48.83	1.12
Sept. 24.	160	3.57	1.85	51.82	.48	Oct. 11	340	3.85	2.13	55.32	.27
Sept. 25.	163	1.97	.94	47.72	.32	Oct. 12.	347	5.01	2.61	52.10	.70
Sept. 25.	167	3.84	1.81	47.65	.73	Oct. 12	353	3.68	2.00	54.35	.53
Sept. 25.	171	3.40	1.77	51.15	.50	Oct. 13.	358	3.41	1.80	52.94	51
Sept. 25.	175	2.95	1.35	45.75	.44	Oct. 15.	362	3.36	1.61	47.92	.50
Sept. 26	181	2.32	1.21	50.80	.38	Oct. 15	366	3.77	1.68	44.58	.72
Sept. 26.	185	3.00	1.87	51.94	.57	Oct. 17.	370	3.30	1.75	53.08	.60
Sept. 27	193	3.03	1.49	49.17	.56	Oct. 17.	375	3.00	1.41	47.00	.50
Sept. 27	197	3.07	2.04	51.38	.64	Oct. 18.	380	4.02	1.05	40.05	.77
Sept. 27.	201	3.95	2.04	51.64	.82	Oct. 18	390	4.00	1.38	36.00	.53
Oct. 1..	210	3.18	1.57	49.37	.42	Oct. 20	397	3.20	1.06	33.12	.61
Oct. 1..	216	2.77	1.49	53.79	.51	Oct. 20	410	2.09	1.22	46.66	.61
Oct. 2..	219	3.10	1.90	56.85	.50	Oct. 21.	417	2.10	.85	40.47	.34
Oct. 2..	222	3.07	1.43	46.56	.59	Oct. 23..	427	4.76	2.13	46.85	.73
Oct. 2..	225	2.63	1.56	59.32	.41	Oct. 23.	431	3.18	1.18	37.36	.57
Oct. 2..	231	1.98	1.01	51.79	.36	Oct. 24.	435	3.40	1.10	35.00	.47
Oct. 3 .	233	3.00	1.96	45.33	.49						
Oct. 3..	237	3.42	1.75	51.17	.48	Mean ..		3.58	1.72	47.72	.57
Oct. 3..	244	3.93	1.98	50.38	.66						

TABLE XXVI.—*Acidity of mill juices.*

[Calculated to malic acid.]

Date.	Fresh chip juice.				Diffusion of Juice.			
	No.	Specific gravity.	N/10 c. c. of NaHO.	Per cent of acid.	No.	Specific gravity.	N/10 c. c. of NaHO.	Per cent of acid.
Oct. 10	318	1.0082	6.9	.17	319	1.0384	4.7	.12
Oct. 10	327	1.0678	9.3	.23	328	1.0351	5.7	.14
Oct. 11	332	1.0709	6.9	.17	333	1.0401	2.2	.09
Oct. 11	337	1.0757	7.8	.17	338	1.0327	3.7	.09
Oct. 12	344	1.0757	8.2	.20	345	1.0302	(*)	(*)
Oct. 12	350	1.0204	7.6	.19	351	1.0204	3.4	.18
Oct. 13	355	1.0613	9.3	.23	356	1.0339	3.3	.08
Oct. 15	359	1.0766	7.7	.19	360	1.0409	4.5	.11
Oct. 15	363	1.0617	7.3	.18	364	1.0351	4.6	.12
Oct. 17	367	1.0717	8.2	.20	368	1.0269	5.6	.14
Oct. 17	372	1.0634	8.1	.20	373	1.0343	4.6	.12
Oct. 18	377	1.0704	5.9	.15	378	1.0347	2.9	.07
Oct. 19	387	1.0726	7.6	.19	388	1.0204	2.7	.07
Oct. 20	394	1.0553	3.1	.08	395	1.0322	2.8	.07
Oct. 20	407	1.0626	4.6	.11	408	1.0322	2.5	.06
Oct. 21	414	1.0557	3.7	.09	415	1.0322	2.2	.06
Oct. 23	424	1.0739	5.0	.12	425	1.0359	2.4	.06
Oct. 23	428	1.0639	5.8	.14	429	1.0209	2.8	.07
Oct. 23	429	1.0290	2.8	.07				
Mean		1.06346	6.6	.17		1.0330	3.6	.09

* Trace.

Table XXVII.—*Fresh chip juice.*

[Comparison of spindle with total solids found by drying.]

Date.	No.	Sucrose.	Glucose.	Total solids by Brix. at 15.5°	Total solids found by drying.	Purity calculated from—		To assist drying.
						Spindle.	Total solids.	
		Per cent.	*Per cent.*		*Per cent.*			
Sept. 29	207	11.72	2.85	18.49	17.38	63.32	67.45	
Sept. 29	207	11.72	2.85	18.49	17.36	63.32	67.51	Asbestos.
Oct. 2 ..	217	11.74	2.33	18.18	16.60	64.57	70.72	
Oct. 2...	217	11.71	2.33	18.18	16.68	64.57	70.38	Do.
Oct. 8...	304	9.76	2.85	16.68	15.49	57.84	63.01	
Oct. 8 ..	304	9.76	2.85	16.84	15.52	57.88	62.88	Do.
Oct. 11..	332	9.32	3.45	17.00	15.24	54.82	61.15	
Oct. 15..	359	11.64	2.65	18.70	17.05	61.71	68.27	
Oct. 19..	387	9.57	3.43	17.87	15.94	54.16	59.88	
Oct. 21..	414	7.37	2.84	13.56	11.91	54.04	61.88	
Mean ..	......	10.43	2.84	17.42	15.92	59.63	65.31	

Table XXVIII.—*Diffusion juice.*

Oct. 1 ..	209	5.80	1.53	9.00	8.14	65.11	71.99	
Oct. 1...	209	5.86	1.53	9.00	8.13	65.11	72.08	Asbestos.
Oct. 3...	236	4.28	1.38	7.66	6.62	55.87	64.65	
Oct. 3...	236	4.28	1.38	7.66	6.63	55.87	64.55	Do.
Oct. 10..	319	5.87	1.44	9.95	9.01	58.90	65.15	
Oct. 12..	345	4.45	1.18	7.58	6.74	54.82	66.02	
Oct. 17..	373	5.31	1.59	9.34	7.90	57.82	66.46	
Oct. 23..	425	5.43	1.75	9.22	8.30	58.91	65.42	
Oct. 24..	433	4.80	1.72	8.34	7.34	57.55	65.89	
Mean ..	......	5.13	1.50	8.64	7.66	58.89	66.86	

WORK DONE AT THE STERLING EXPERIMENT STATION.

REPORT OF A. A. DENTON AND C. A. CRAMPTON.

The experimental work which has been done at the Sterling Sugar Experiment Station was wholly in the line of improving the sorghum plant with a view to increase the yield of sugar from sorghum canes, to obviate certain physical or outward faults of the plant, and to obtain varieties which are less variable in their yield of sugar.

It is probable that the extraction of juice from sorghum canes has nearly or quite reached its practical limit, and that diffusion apparatus needs only to be improved in details of construction which is more properly the work of machinists.

It is probable that the evaporating apparatus used in sugar manufacture, the triple effect, the vacuum pan, etc., will not soon be very greatly improved, for they are the result of many years of experiment by scientists, aided by the most skilled engineers.

There remains, however, a very important and promising field for experimental work in the line of sugar manufacture, and that is the improvement of the sorghum plant upon which the sorghum-sugar industry depends for ultimate success.

The importance and necessity of such work has been recognized by every one who has been engaged in the development of the industry, but very little has been actually done in that direction ; the greatest attention has been devoted to the methods of extraction and manufacture, while the quality of the raw material has been neglected.

If improved varieties of sorghum were developed, as improved varieties of the sugar-cane or of the sugar-beet have been developed, a successful future for the sorghum-sugar industry in competition with the sugar-cane and the sugar-beet industries could be confidently assured.

In illustration of this disability which hinders the sorghum-sugar industry, it is proper to recall the fact that the new beet-sugar factories erected this year in California imported beet seed from Europe at heavy cost, because there the sugar-beet has been bred up and improved by many years of persistent effort by experts in that line, so that this European improved beet seed produces at once in California beets which contain from 14 to 20 per cent. of sugar. New sorghum-sugar factories have been built this season in Kansas, but they can nowhere procure similar improved sorghum seed, for the sorghum plant has yet to be developed and improved. As an instance of the necessity for the exercise of care in the selection of seed, the experience of two of the new factories this season may be cited. One of us visited the factories at Douglass and Conway Springs at the beginning of the season, about September 7. At the latter place there was great complaint of the quality of the early cane; seed had been obtained, supposed to be pure Early Amber, but seed of later varieties, such as Orange, had been allowed to become mixed with it in considerable quantities, and the result was a field of cane of which the greater part was fully ripe and ready for working, while a portion was still green, with the seed not yet out of the dough. It required entirely too much labor to separate it in the field, and when the cane was cut and brought to the factory the green cane lowered the average of the whole to such an extent that it was hardly fit to work for sugar. At Douglass about 100 acres had been planted for early cane, with seed supposed to be Early Amber. As the factory was greatly delayed in starting up, fears had been entertained that this cane was overripe and deteriorating. Examination showed this "early cane" to be not Early Amber at all, but the old-fashioned Chinese, a variety which, with us at least, did not attain its maximum of sugar content until quite late in the season. Had the factory gotten into operation by the middle of August, as they expected, they would have found their "early cane" entirely too green to make sugar.

THE ORIGIN OF THE EXPERIMENTAL WORK AT THE STERLING SUGAR EXPERIMENT STATION.

In the spring of 1888 the Sterling Sirup Works planted all the varieties of sorghum which, with the time and means at their command, they could procure in this or in foreign countries, in an experimental field, under as similar conditions as possible, in order to enable them to compare the qualities of the canes of the numerous varieties, with a view to selecting the best varieties for future cultivation. They had in mind a similar experimental plantation in Jamaica, where sixty to seventy varieties of the sugar-cane have for many years been grown in order to select the varieties which were best suited to the West Indies,* the result of which is shown by the fact that an improved variety of sugar-cane, which is sometimes called "Jamaican," because it was grown at and introduced by the Jamaica experimental station, is now giving an extraordinary yield of sugar in many places.

They were induced to undertake this experimental work by the necessities of their business. In the past seven years they have produced, each year, from 500 to 700 acres of cane, and have manufactured the

---

* Analyses of samples of these different varieties from a collection exhibited at the New Orleans Exposition in 1885 were made by C. A. Crampton, at the Sugar Laboratory of the Department of Agriculture, in its exhibit. The results of these analyses were published by Prof. Morris in the *Jamaica Official Gazette*.

crop. Each year they have planted the common varieties, and also varieties new to them which they could readily procure. The selection of better varieties and the improvement of the quality of the canes is a matter of importance to them, as it is to all others who are concerned in the sorghum industry.

It appeared to the Sterling Sirup Works that the first step to be taken in improving the sorghum plant was to collect as many varieties as possible, from all localities where sorghum is grown, to acclimate them, and to practically test the numerous varieties in all the points which constitute a good variety of sorghum.

It is now to be regretted that a much more extended search was not made, in this and in foreign countries, for other rare and unknown varieties, but they then regarded this year's work as only the beginning of a private research which would continue for some years.

The object of the experimental work was to improve the sorghum plant.

(1) Improved varieties of sorghum should be developed, producing canes of uniform saccharine quality, to lessen the unusual variableness which now characterizes the sorghum plant.

(2) The physical or outward character of the canes should be improved to obviate faults and also to increase the yield of cane in tons per acre.

(3) The percentage of cane sugar in the juices of the cane should be increased.

(4) The percentage of substances in the juice which lessen the yield of sugar should be diminished.

THE NECESSITY FOR IMPROVING THE SORGHUM PLANT.

The sorghum plant is adapted to large areas of the country which are not adapted to the production of sugar from the sugar cane or from the sugar beet. It is especially adapted to the dry climate of the great West. Its cultivation is suited to the habits of the farming population. When the sorghum plant has been successfully developed and improved as other sugar-producing plants have been improved, the sorghum-sugar industry will prosper and will employ capital and labor in producing the sugar which we now import.

THE FAULTS OF THE SORGHUM PLANT.

The sorghum plant is sometimes a good sugar-producing plant, sometimes it is merely a sirup-producing plant. This variability in the chemical composition of its juices is what might be expected from a plant which has not yet been bred up to fixed types of excellence by long-continued selections of seed from the finest plants of the best varieties.

In this connection it is interesting to note that in 1747 the chemist Marggraff was able to extract 5 per cent. of sugar from the beet; fifty years afterwards the chemist Achard was able to extract but 1 per cent. of sugar, and the eminent chemist Sir Humphrey Davy published positive assertions that beet sugar could not be made profitably, an l that it was not fit for use. Sixty-five years after Marggraff had ex- tracted 5 per cent. of sugar from the beet the beet-sugar factories realized only 2 per cent. of sugar from it. These facts seem to indicate that the sugar beet was variable until the plant had been developed.

Besides the variability of the sorghum plant there are other faults which pertain in greater or less degree to the different varieties. Some varieties are long and slender reeds with heavy seed tops and the canes are liable to lodge and tangle in storms. This fault greatly increases the difficulty of harvesting the canes, and the "down" or lodged canes are also inferior in saccharine value.*

Some varieties "tiller;" that is, one root produces several canes, just as one grain of wheat produces several stalks. It is injurious because the secondary canes ripen at different periods, and in harvesting large fields of cane it is impossible to avoid mixing overripe, ripe, and unripe canes. Some varieties have a habit of producing false or secondary seed-heads. As soon as the cane approaches maturity, and often before that period, it forms two or more new seed-heads, which rapidly de- velop. This delays the ripening of the cane and lessens the yield of sugar. Some varieties, as soon as fully mature, produce offshoots from each joint of the canes and also offshoots from the roots, and the sugar in such rapidly disappears. Some varieties rapidly deteriorate in the quality of the juice as soon as they are ripe, and allow little time to manufacture the canes. Some varieties mature very small seeds, and these produce plants which are weak and slow-growing in the first weeks of their existence and are kept clear from the more vigorous weeds with greater difficulty than the stronger plants, which are pro- duced by larger seeds. Some varieties have very impure juice and some have strongly acid juice. Some varieties give light yield of cane, light yield of juice, and light yield of seed. Some varieties obstinately retain the glume or envelope of the seed grains, so that it can not well be separated by ordinary means. Analyses seem to show that the clean grain of sorghum seed is practically equal in value to corn as food for stock, but the adhering glume or envelope contains tannin, which is injurious; and some varieties contain much of this substance and some but little. Some varieties mature so late that they give but little time to manufacture the canes before frost.

---

* This deterioration of lodged cane has been often noted before, but the following analysis, made at this station, may serve to emphasize it:

	Solids.	Sucrose.	Glucose.
	Per cent.	Per cent.	Per cent
Average sample standing cane	10.49	11.02	2.03
Same of down cane of the same plat, only sound stalks taken	13.29	6.76	3.57

THE FAULTS OF THE SORGHUM PLANT AND OF THE SUGAR BEET COMPARED.

The sugar beet contains mineral substance which lessens the yield of sugar. As a rule these mineral substances in the juice vary indirectly as the sugar varies; that is, the greater the percentage of sugar the lower the percentage of mineral substance.

Sorghum contains glucose in the juice, and this lessens the yield of sugar. As a rule the percentage of glucose in the juice varies inversely as the percentage of sugar varies, that is, the greater the percentage of sugar the less the percentage of glucose.

The beet has also physical or outward faults. It is a biennial plant; it stores sugar the first season, it produces seed the second season. Sorghum is an annual plant; it produces sugar and also seed in one season; but when it has produced its sugar and its seed it often attempts a second crop of seed, and this lessens the yield of sugar.

The sugar beet sometimes makes a "second growth." Sorghum sometimes sends out offshoots from every joint and offshoots from the roots.

The sugar-beet is sometimes hollow. Sorghum canes are sometimes pithy and contain but little juice.

The sugar-beet is sometimes attacked by the "brown penetration," a discoloration of the beet which lessens the yield of sugar. Sorghum canes sometimes have brown or red spots in the interior of the canes. The sugar-beet often had faults of form; it had forked roots, making harvesting the beets and cleaning them from dirt more difficult. Sorghum also has faults of form.

CAN THE SORGHUM PLANT BE IMPROVED?

Judging by all analogies, the sorghum plant can be very greatly improved by intelligent and long-continued selection. Stirpiculture in the animal kingdom has given us the Cotswold sheep, the Poland-China hog, the Jersey cow, and the Norman horse. In the vegetable kingdom it has given us the Peabody corn, the Zinfandel grape, the Lapice sugar-cane, and the Klein-Wanzleben sugar-beet. It has been truly said, "Wherever and with whatever plant selection of the best for seed has been long continued wonderful results have been obtained." Darwin said, "Let any common plant, even a roadside weed, for instance, be grown on a large scale, and let a sharp-sighted gardener select and propagate slight variations, and see if new varieties do not result." Knauer started with a variety of the sugar-beet which contained but 11 per cent. of sugar; he improved it by selecting the best for seed until he produced the "Imperial" variety, which contained 16 per cent. of sugar. Deprez et Fils, by selection of seed from the best roots, produced three varieties which contained from 14 to 16 per cent. of sugar. Vilmorin, the celebrated horticulturist of France, created the "Improved Vilmorin," improved in form and in yield of sugar. There are no apparent reasons why the sorghum plant may not be improved by diligent use of similar methods.

THE METHODS OF IMPROVING THE PLANT.

The principal methods of improving the plant may be stated as follows:

(1) By growing and testing all known varieties and selecting the most promising.
(2) By hybridizing or crossing these varieties.
(3) By preserving "sports" or variations.
(4) By selecting seed from the finest individual canes of each variety.
(5) By improved methods of cultivation.

All of these methods have been practiced to a greater or less extent in the work at this station, and the results will be set forth in the order given above. It must be remembered, however, that the results accomplished in this direction by one season's work can be at best but a mere beginning. To attain the end desired in the improvement of the plant the continuation of such work over a series of years is indispensable. If this season's work and the methods pursued will serve to point out the necessity and importance of this line of investigation, and, in general, the manner in which it may be best carried out, a great deal will have been accomplished. It is hardly necessary to call attention to the desirability of following up the system of development thus opened up; and it is to be hoped that opportunity may be afforded the Department in the future to carry on this work, which promises to be of the greatest value to the sorghum industry.

I. EXPERIMENTS IN GROWING DIFFERENT VARIETIES OF CANE.

It is probable that all varieties of sorghum are not equally well adapted to all localities where sorghum is grown.

Some varieties have peculiarities which cause them to succeed best in certain places. The Early Amber, for instance, probably succeeds better and has more valuable qualities in Iowa than in Texas.

There is an analogy in this with other plants. A Rhenish variety of the grape succeeds best in dry soil. A Swiss variety succeeds best in wet climates. Spanish varieties of wheat do not succeed in Germany. English wheat does not thrive in India.

To select the best varieties of sorghum for a given locality it is necessary to grow all known varieties there and to select those which prosper best under its conditions.

It is not now easy to collect seed of numerous varieties of sorghum. The common varieties only are for sale by seed dealers; other varieties can only be found among distant cane-growers in this and in foreign countries. In collecting many varieties, duplicates of some varieties are obtained, because a single variety often has many names. This is natural in foreign countries, where different languages are used; but in our own country the same variety often has many names, which are usually derived from some peculiarity of the plant. This is also true of other plants. It is said that all the varieties of the sugar-beet may be classed in four groups; there seem to be twenty-three principal varieties, which have several hundred names.

The varieties of sorghum often can not be distinguished by the appearance of the seed alone, or even by the seed-heads alone. They can best be classed by observing the growing canes. Varieties which have long been grown under very different conditions often vary enough from the usual type to be classed as subvarieties. The Chinese cane from Australia differs in some respects from the Chinese from Central America, and that differs in some respects from the Chinese of this country.

These facts add to the difficulty of classifying the numerous varieties of sorghum. Sorghum is also grown in opposite hemispheres, and the proper season to collect varieties in one country is not the proper season in another country.

ACCLIMATIZATION OF VARIETIES.

In growing and comparing varieties of sorghum which have been obtained from different localities it is necessary to consider acclimatization. Plants, as well as men and animals, require time to adjust themselves to new conditions. Linnæus said seed from tobacco grown in Sweden ripened a month earlier than that from foreign seed. Seed corn taken from Virginia to New England ripens with difficulty the first season. Seed corn taken from New England to latitude 45° ripens with difficulty the first season. In both cases they mature perfectly in their new homes after a few seasons. European plants produced in India, from seeds grown in India a few seasons, succeed better than from directly imported seeds.

These facts indicate that it is necessary to acclimate varieties of sorghum procured from localities which have different conditions of soil and climate before correct comparisons of their qualities can be made.

In the experimental work at this station it was especially noticeable that varieties of sorghum received from localities having long and warm growing seasons produced larger canes, which matured later than the same varieties from Northern localities. It was also noticeable that varieties received from localities which have little rain-fall succeeded better this season than the same varieties received from localities having excessive rain-fall.

It is obvious that a plant may be removed many thousand miles with slight change of environment and that it may be easily acclimated. It is also obvious that a plant may be removed a comparatively short distance with considerable change of conditions and may be acclimated with difficulty. It seems to require three years to acclimate varieties of sorghum which have been grown under very different conditions. These facts require consideration when making selections from numerous varieties grown the first time in this country.

VARIETIES GROWN AT THE STERLING EXPERIMENT STATION.

There were about 250 different plots of sorghum grown at this station; of these 150 were crosses, selected by Mr. Denton; the remaining 100 plots were planted with varieties presumably distinct, though more than one plot was planted of a few standard varieties from seed obtained from different localities. Of those supposed to be distinct varieties, however, though sent in under different names, many were found to be duplicates, showing minor variation perhaps, but not sufficient to entitle them to classification as distinct varieties.

For instance, seeds of the well-known variety Red Liberian were received bearing the names "African," "Sumac," "Clubhead," "Rio Blanco," etc.; samples of Honduras seed were named "Honey cane," "Broom cane," "Silver top;" samples of Chinese cane seed were received as "New sugar-cane," and "Sorghum saccharatum." It will be seen in the following analyses that seeds of the same varieties received from different localities produced canes of quite different qualities.

Thirty-six of the varieties proved to be non-saccharine, useful for forage purposes, but not containing enough saccharine matter to be of value as sugar-producing plants.*

In addition to most of the varieties grown in the United States the list includes many obtained from Asia, Africa, and South America. The seeds of many foreign varieties were injured by dampness and by insects; of some of these not a single seed germinated.

The experimental lots varied in size, some containing a few acres of each variety and some containing 150 hills of cane, and of some varieties but a few canes were grown. In the experimental grounds of the Jamaica Botanical Gardens, where sixty to seventy varieties of sugar-cane are grown, each variety occupies from one-half to one-third of an acre. This seems to be the proper size for experimental lots, as it allows selections of seed to be made from the best canes of each variety. Moreover, foreign seeds often germinate poorly, and when small plots are planted there is often not a single cane of some varieties produced, as was the case here.

COMPARISON OF THE VARIETIES BY ANALYSIS.

It is not an easy matter, as might seem to be the case at first sight, to make a comparison of different varieties by the analysis of juices from selected samples. In the first place, to make a fair comparison between varieties they should be taken at their maximum of maturity, and this is a point which can not be determined by any outward sign, but only by actual analysis. Then the difficulties of sampling can only be properly appreciated by one who is familiar with them. Add to these the difficulties of comparison, the obstacles in the way of always getting uniform conditions in the growth of the plots themselves, attacks of

---

* Of the non-saccharine varieties, 20 were derived from China, 8 from Africa, 3 from India, and 5 from this country; the seed from all these were carefully preserved, and will be distributed by the Department. Many will doubtless prove new and valuable acquisitions as forage plants.

chinch bugs in one plot and not in another, a sandy spot in one and not in another, imperfect germination of seed in one plot causing a thin stand, while in other plots the canes stand close together, and it will be seen that the task of differentiating between varieties by growing them in plots and submitting the canes produced to analysis is by no means an easy one. It is a very complex problem. One season's work should never be held conclusive; a variety may have been placed at a disadvantage from some one of numerous possible causes.

In the work here the varieties were analyzed as often as possible, to avoid the error of having analyses of either unripe or overripe canes to compare with the analyses of other varieties at their maximum; the highest analysis in the series may be taken as the basis of comparison. The error of sampling was avoided as much as possible by taking good-sized samples, and by having them all taken by one and the same person.* The errors arising from differences of growth were augmented, unfortunately, by irregularities in the time of planting; some lots of seed being received very late in the spring. The time of planting is noted with each plot.

EARLY VARIETIES.

Several of these gave very satisfactory results, so far as early ripening was concerned. The late date at which the laboratory was established at the station did not admit of many analyses before they had passed their maximum of maturity. In the case of the Early Tennessee and Whiting's Early Variety this point had probably been passed before any analysis was made.

Plot.	Date.	No. of analysis.	Degree Brix.	Sucrose.	Glucose.	Co-efficient of purity.	Remarks.
				Per cent.	Per cent.		Swain's Early Golden, planted May 8, produced from seed which was ten years old; the canes were good; unmixed.
93.....	Aug. 24	3	18.03	12.88	1.92	71.44	
	Sept. 1	....	16.09	11.24	2.04	69.86	
	Sept. 10	134	15.90	10.53	1.54	65.85	
102.....	Aug. 24	6	15.54	8.45	2.55	54.38	Early Tennessee, planted May 8, matures at least ten days earlier than Amber. The canes are small, but it is worthy of further trial on account of its earliness.
	Aug. 31	56	13.08	6.44	1.90	49.24	
233.....	Aug. 25	12	16.64	10.42	2.89	62.62	Whiting's early variety, planted May 19. This originated in New York from a single cane, which was the only one which ripened before frost in a field of Early Amber. The canes are small and ripen two weeks earlier than Early Amber. In these lots several canes showed reversion to the Early Amber type, which seems to indicate that this variety was formed by an accidental cross, and that a cross may be earlier than either parent.
	Sept. 6	113	15.72	9.66	1.96	61.45	
	Sept. 8	126	14.52	7.43	2.08	51.17	
234.......	Aug. 30	46	15.63	10.30	1.48	65.90	Another plot of the same.

* Mr. Denton did all the sampling himself.

EARLY AMBER.

This is the most widely known of all the varieties of sorghum. It was included in almost every collection of seeds obtained in this or in foreign countries. It was obtained from Australia, from Algeria, and from South Africa, which shows its wide distribution.

It is an excellent variety for sirup and for sirup-making if the canes are cut when in their best condition. The juice is then comparatively pure and has a pleasant taste. It deteriorates rapidly in this climate soon after it matures. This is a serious fault in sugar manufacture where very large fields of this variety are grown. It also yields less weight of cane and less seed than many other varieties. It will, however, retain a place in the list of varieties planted for sugar manufacture. It probably succeeds better in Minnesota and Iowa than in Louisiana and Texas.

There are several subvarieties of Early Amber; the Black Amber, the White Amber, the Golden Sirup, the Cape May Hybrid, etc. The early varieties given above might be considered subvarieties of the Early Amber, as they were undoubtedly derived from it.

In the experimental field Early Amber was planted at intervals from May 5 to July 5, and one lot ripened after another, prolonging the time for analyzing the canes. Seeds of Early Amber received from widely different localities were planted to compare the qualities and to observe differences. Amber was also often planted near lots of unknown varieties to assist in determining their time of maturity. It is evident that many circumstances affect and vary the time of maturing of a new variety. By comparing it with a well-known variety such as the Amber, planted under the same conditions, its time of maturing in any season can be accurately determined.

The first analysis was made August 24, at which time the earliest plantings were spoiled and some of the later were deteriorating.

Plot.	Date.	No. of analysis.	Degree Brix.	Sucrose.	Glucose.	Coefficent of purity.	Remarks.
				Per cent.	*Per cent.*		
90........	Sept. 7	120	12. 88	6. 50	3. 14	50. 47	Black Amber, planted May 8, produced fine canes, which ripened a week earlier than Early Amber. It was overripe when samples were taken for analysis.
92..... {	Aug. 24	....	18. 03	12. 99	1. 84	72. 05	White Amber, planted May 8, produced fine canes, pure and unmixed, with but slight differences from Early Amber.
	Sept. 1	64	17. 20	12. 21	1. 78	70. 62	
	Sept. 10	133	14. 80	9. 20	1. 72	62. 16	
237 .... {	Aug. 31	48	13. 61	7. 95	3. 16	58. 41	Early Amber, planted May 22; seed received from Sydney, Australia.
	Sept. 8	128	14. 40	9. 01	2. 62	62. 57	
	Sept. 17	223	12. 76	8. 32	1. 86	65. 20	
	Sept. 24	317	14. 63	8. 44	2. 92	57. 69	
1........	Aug. 25	8	17. 54	13. 18	1. 07	75. 14	Early Amber, from seed which had been grown by the Sterling Sirup Works at this place for six years; planted May 5; produced large canes.

Plot.	Date.	No. of analysis.	Degree Brix.	Sucrose.	Glucose.	Coefficient of purity.	Remarks.
				Per cent.	*Per cent.*		
211	Aug. 30	43	15.17	9.39	2.83	61.90	Early Amber, planted May 21, received from Cape Town, South Africa, labeled "Holcus Saccharatus."
	Sept. 8	129	12.99	6.56	2.82	50.89	
	Sept. 26	340	12.00	6.11	2.09	50.92	
23	Aug. 24	4	18.10	13.70	1.12	75.69	Early Amber, from New York, planted May 6, produced large and handsome canes.
248	Aug. 31	49	15.33	10.80	2.10	70.22	Folger's Early variety, planted May 24. It originated in continuous selections, made in Iowa, from Early Amber; produced good canes.
249	Sept. 8	125	16.02	10.91	1.54	68.10	Another plot of the same.
	Sept. 17	224	16.16	11.43	1.09	70.73	
	Sept. 26	347	16.10	11.48	1.07	70.86	
	Oct. 4	400	16.78	11.94	1.40	71.16	
	Oct. 9	539	15.02	10.25	2.28	65.02	
	Oct. 15	500	14.68	9.94	1.51	67.85	
	Oct. 19	630	15.81	10.99	1.31	69.51	
	Oct. 22	658	15.41	10.22	1.82	66.32	

The following lots of "Chinese" cane showed interesting differences in habit, owing probably to conditions in which they had been previously grown:

Plot.	Date.	No. of analysis.	Degree Brix.	Sucrose.	Glucose.	Coefficient of purity.	Remarks.
				Per cent.	*Per cent.*		
62	Aug. 29	28	13.94	6.02	4.07	43.19	Chinese, received as "the New Sugar-cane," from Central America, planted May 7.
	Sept. 7	123	15.27	8.44	2.96	55.27	
	Sept. 17	214	16.20	9.27	2.71	57.22	
	Sept. 26	355	15.84	9.41	2.00	59.41	
	Oct. 6	488	17.18	11.42	1.84	66.47	
	Oct. 8	510	16.00	9.48	2.16	59.25	
	Oct. 15	593	17.37	11.79	1.35	67.88	
216	Aug. 30	41	14.10	6.51	3.67	46.17	Another lot of same, planted May 18.
	Sept. 18	233	15.00	7.90	2.45	52.67	
	Sept. 29	406	16.56	9.99	2.28	59.78	
	Oct. 8	512	16.50	10.49	2.09	63.58	
	Oct. 13	592	15.34	8.97	2.04	58.47	
215	Aug. 30	42	14.10	7.55	1.51	53.55	Chinese, received as "Sorghum saccharatum," from New South Wales, planted May 19. But few seeds grew and canes were poor.
	Sept. 17	216	16.29	9.76	2.25	59.91	
	Sept. 29	405	16.06	9.43	2.08	58.72	
	Oct. 8	509	15.11	8.83	2.19	58.44	
232	Aug. 30	44	15.10	7.13	4.19	47.22	Chinese, received as "Sorghum saccharatum," from Cape Town, Africa, planted May 15. Good canes.
	Sept. 17	219	15.35	8.02	2.48	56.16	
	Sept. 28	395	18.20	11.57	2.19	63.57	
	Oct. 3	445	16.49	11.02	2.03	66.83	
	Oct. 8	508	17.98	12.46	1.44	69.30	
	Oct. 15	591	17.22	11.20	1.83	65.39	
37	Aug. 25	9	14.04	7.71	3.66	51.61	Chinese, grown in the United States, planted May 5; produced better canes than the foregoing lots, which seems to indicate that it has become adapted to this country.
	Sept. 10	140	17.41	10.43	2.40	59.01	
	Sept. 11	148	18.52	12.31	2.08	66.47	
	Sept. 17	213	17.45	10.83	2.16	62.06	
	Sept. 26	348	18.05	12.95	1.67	69.44	
	Oct. 8	511	19.00	13.23	1.40	69.63	
69	Sept. 19	255	15.30	9.96	1.66	65.10	White India, planted May 8; produces large and handsome canes, free from offshoots; yields seed well. The seed did not seem to germinate well and the stand was poor.
	Sept. 26	360	16.16	10.18	1.08	62.99	
	Oct. 5	476	15.50	10.53	1.12	67.04	
	Oct. 9	531	17.07	13.07	1.03	73.06	

Plot.	Date.	No. of analysis.	Degree Brix.	Sucrose.	Glucose.	Coefficient of purity.	Remarks.
				Per cent.	Per cent.		This lot was planted with seeds labeled "Enyama," from Louisiana; planted May 8. The canes are very similar to, and probably identical with, the White India.
70	Sept. 19	261	15.82	10.09	2.04	63.78	
	Sept. 26	359	19.05	13.26	1.61	60.61	
	Oct. 5	477	18.10	12.74	1.47	70.39	
	Oct. 9	530	16.70	11.61	1.59	69.52	
67.....	Sept. 19	256	16.92	12 04	1.47	71.15	White Mammoth.* planted May 8. There is apparently no difference between these canes and the White India or the Enyama.
	Sept. 26	353	16.46	11.14	1.44	67.68	
	Oct. 5	474	16.33	11.00	1.26	72.87	
	Oct. 9	533	16.95	11.50	1.24	67.85	

* The botanical description of the White Mammoth is (Ann. Rep. U. S. Dept. of Agriculture, 1880, 40) " Heads very dense, expanding toward the flattened top: glumes shining, black, prominent; seed white, large hilum, inconspicuous." This applies well to the White India, Enyama, and White Mammoth, as grown in the above lots, but the seeds may have been incorrectly named by those who sent them.

VARIETIES OF ORANGE.

A large number of subvarieties of this standard variety exist, but the differences in character are probably less than with other varieties, considering the opportunities that have been offered it for variation ; that is to say, the predominant race characteristics hold their own better throughout the crosses. Doubtless this is due to the fact that it has been grown a long time and become well established. Some of the plots of Early Orange showed remarkable uniformity, not a single important variation being found in them. This stable quality will make the variety very useful in crossing where certain stable qualities are desired.

Plot.	Date.	No. of analysis.	Degree Brix.	Sucrose.	Glucose.	Coefficient of purity.	Remarks.
				Per cent.	Per cent.		
236.....	Sept. 3	87	12.55	5.35	4.48	42.62	Early Orange, received from Department of Agriculture; planted May 22. Mixed and irregular canes.
	Sept. 5	97	13.78	7.51	3.58	54.50	
	Sept. 17	222	15.60	9.25	3.61	50.20	
	Sept. 24	316	14.63	8.60	2.90	58.78	
	Sept. 25	342	17.12	11.29	2.70	65.95	
81	Sept. 27	371	17.58	12.82	1.33	72.02	Early Orange, from Fort Scott, Kans.; planted May 8.
83*.....	Sept. 2	161	16.66	10.49	2.80	62.96	Early Orange, from South Carolina; planted May 8.
	Sept. 19	269	16.53	10.02	2.64	60.62	
	Sept. 27	370	16.88	10.52	2.30	62.32	
	Oct. 9	538	17.59	11.53	2.74	65.55	
	Oct. 19	631	15.52	9.54	2.50	61.47	
87........	Sept. 19	270	16.53	11.39	2.26	68.91	Early Orange, from Arkansas; planted May 8.
68......	Sept. 12	157	17.80	11.37	2.70	63.88	Received from Louisiana as White Mammoth, but produced Early Orange canes; planted May 8.
	Sept. 19	259	16.77	10.53	1.96	62.79	
	Sept. 25	337	17.77	10.98	3.05	61.79	
	Oct. 9	532	17.77	12.87	1.55	72.43	
	Oct. 15	601	17.90	12.90	1.13	72.07	
82	Sept. 19	267	15.10	9.53	2 14	63.11	Kansas Orange; planted May 8. Strong and stocky canes.
	Sept. 27	369	14.88	9.23	1.77	62.03	
49......	Aug. 29	25	11.85	6.02	3.23	50.80	Same; planted May 8.
	Sept. 11	145	16.20	10.81	1.89	66.73	
	Sept. 18	247	15.57	10.50	1.71	67.44	
	Sept. 26	354	18.04	12.19	1.62	67.57	
	Oct. 9	523	16.91	12.17	1.20	71.97	
88......	Sept. 12	156	14.02	5.97	4.16	42.56	New Orange; planted May 8. Plot was injured by chinch-bugs.
	Sept. 19	271	14.12	6.47	8.06	45.82	
	Sept. 27	375	16.25	9.53	3.07	58.05	

* Plot 228, same as 83; planted May 21. See experiments in development, page 123. (This plot was very uniform, not a single variation being found in it.)

Plot.	Date.	No. of analysis.	Degree Brix.	Sucrose.	Glucose.	Coefficient of purity.	Remarks.
				Per cent.	*Per cent.*		Late Orange, from New
59	Sept. 13	158	16.12	0.94	2.41	61.66	Jersey; planted May
	Sept. 19	274	17.02	10.58	3.07	62.16	8; produced large and
	Oct. 9	540	17.90	12.73	2.32	70.76	strong canes, which re-
	Oct. 15	600	16.97	11.43	2.30	67.35	mained long in good condition.
	Aug. 20	29	10.48	3.15	4.32	30.06	
	Sept. 11	151	16.82	10.31	2.80	61.30	
	Sept. 19	266	17.18	11.06	2.75	64.38	Received as "Early
	Sept. 27	364	18.85	13.38	1.83	70.98	Gooseneck;" planted
75......	Oct. 5	473	17.48	12.48	1.58	71.40	May 8. The canes
	Oct. 9	537	18.32	12.79	1.46	69.81	were apparently iden-
	Oct. 15	602	15.70	10.22	1.98	65.10	tical with the Late Or-
	Oct. 19	649	17.60	12.63	1.03	71.76	ange.
	Oct. 12	663	16.81	11.15	1.78	66.33	
	Oct. 24	683	16.10	9.72	2.72	60.37	
							Medium Orange; said to
	Aug. 29	30	15.90	10.48	1.00	65.91	be a cross between
81......	Sept. 7	122	16.70	11.35	1.07	67.96	Early Amber and Kan-
	Sept. 12	160	15.56	9.91	1.24	63.00	sas Orange; received from Illinois; canes
							much resemble the Amber; planted May 8.
	Aug. 31	47	11.81	5.92	2.48	50.13	
235	Sept. 8	127	11.48	9.22	1.59	63.07	Same; planted May 22.
	Sept. 17	218	11.83	9.67	1.28	65.12	
	Sept. 26	311	16.70	11.84	1.04	70.30	

RED LIBERIAN.

This old variety, known under so many synonyms, made a very good showing in this season's trial. It is very distinctive in its character, gives a good yield per acre, and has a good content of sugar. Its greatest fault is to be found in the small round seeds it produces. These produce plants which are very small and feeble at first, and when the planting is done with a corn-planter the seed is apt to be too close in the hill.

Plot.	Date.	No. of analysis.	Degree Brix.	Sucrose.	Glucose.	Coefficient purity.	Remarks.
				Per cent.	*Per cent.*		
	Sept. 19	202	16.42	9.09	3.68	55.36	Red Liberian, received
72.....	Sept. 27	363	18.27	12.15	2.73	66.50	from Missouri as "Lit-
	Oct. 4	402	18.80	13.25	2.74	70.48	tle Sumac;" planted
	Oct. 5	470	18.00	12.45	2.40	69.17	May 8; large yield of
	Oct. 9	514	18.25	12.07	2.94	66.14	good canes.
	Sept. 19	263	19.91	14.26	1.90	71.62	Red Liberian, received
73......	Sept. 27	367	19.92	14.76	1.84	74.10	from Texas under the
	Oct. 5	471	18.80	13.52	1.67	71.91	name of "Red Top;"
	Oct. 9	535	19.45	13.90	1.71	71.47	planted May 8.
	Sept. 17	225	16.66	12.34	2.51	74.07	
	Sept. 19	264	17.38	9.69	4.11	55.75	Red Liberian, received
74	Sept. 27	366	18.82	11.64	3.24	61.85	under the name of
	Oct. 5	472	18.28	11.71	3.16	64.06	"Gooseneck;" planted
	Oct. 9	686	20.25	13.90	2.31	68.15	May 8.
	Oct. 30	692	16.70	4.89	4.94	31.15	
222.....	Sept. 28	396	17.66	11.12	3.22	62.97	Same as No. 74; planted
	Oct. 5	481	19.20	13.56	1.42	70.03	May 18.
							Golden Rod, from Georgia;
							planted May 8. The
95........	Sept. 5	99	14.35	6.96	4.48	48.50	seed planted in the plot did not germinate, and a sample was taken from a field of the Sirup Works. It was probably overripe.

118

Plot.	Date.	No. of analysis.	Degree Brix.	Sucrose.	Glucose.	Coefficient of purity.	Remarks.
99......	Aug. 31	.61	14.52	5.57	3.78	38.30	Honey Dew, from Indiana; planted May 8; produced fine canes; yields white, clean seed in large quantities. This plot did not have favorable conditions; a large field planted by the Sirup Works gave better results. The last analysis given was taken from this field, not from the plot.
	Sept. 7	118	14.77	8.36	1.19	56.60	
	Sept. 11	153	11.98	4.99	1.69	41.65	
	Sept. 5	103	17.53	11.34	1.62	64.69	
250*....	Oct. 4	457	16.60	10.78	1.33	64.94	Same; planted May 18. This plot did better than the earlier planting.
	Oct. 10	613	15.75	9.34	1.64	59.30	
	Oct. 22	660	18.10	11.92	2.02	65.86	
98......	Aug. 26	11	15.84	8.29	4.27	52.31	Dutcher's Hybrid; originated in Iowa; planted May 8.
	Sept. 3	85	16.24	10.50	2.14	64.06	
223.....	Aug. 28	19	12.74	5.56	3.86	43.64	Same; planted May 18.
	Aug. 28	20	13.15	6.10	3.73	46.39	
	Sept. 18	229	16.50	10.05	2.28	60.91	
97........	Aug. 31	58	12.00	5.76	2.55	48.00	Link's Hybrid; planted May 8. This plot with the New Orange, and several others near this, were injured by drought and insects so that analysis were not continued on it. A large series of analyses were made on another plot of the same variety, which are given under the experiments in development, page 122.
101.....	Aug. 25	10	12.42	4.57	4.70	36.79	Price's Hybrid; said to be a cross between Amber and Honduras; planted May 8.
	Sept. 3	84	16.84	10.77	2.71	63.95	
	Sept. 6	111	16.10	9.20	2.96	57.14	
214.....	Sept. 18	231	17.00	10.81	1.79	63.59	The Planter's Friend; received from Australia; planted May 18. This variety was tested at the Government farms in Madras, India, in 1882, with two other varieties, and was considered the best in saccharine qualities. It is a promising variety.
	Sept. 30	418	20.50	13.83	1.66	67.46	
	Oct. 4	455	18.06	10.96	3.38	60.68	

* Plot 100. Honey Drip, from Texas; planted May 8. Large stocky canes and large seed-heads. Analyses given under development experiments, page 123.

HONDURAS.

This variety is widely known and distributed under its various names of Sprangle-top, Broom-cane, etc. It produces a larger yield per acre than any other well-known variety of sorghum. It has been known at Sterling to yield as high as 33 tons of field cane per acre. In none of the plots planted with it did it ripen sufficiently to show maturity. The hope for this variety lies in its improvement by selections of early maturing canes.

Plat.	Date.	No. of analysis.	Degree Brix.	Sucrose.	Glucose.	Coefficient of purity.	Remarks.
				Per cent.	Per cent.		Honduras from Louisiana, very large canes but more affected by drought than No. 64.
64 ..... {	Oct. 6	483	15.75	9.34	3.35	59.30	
	Oct. 15	593	15.51	9.54	3.24	61.39	
63	May 8	....	....	....	....	....	{ Honduras, from Arizona. No analysis was made.
65	May 8	....	....	....	....	....	Honey cane from Texas. Slight differences from Honduras. No analysis made.
66 ..... {	Sept. 19	254	14.92	8.35	4.19	55.96	"Silver Top or Broom Cane" from Texas; planted May 8. The handsomest lot of the Honduras canes; large weight of cane did not ripen.
	Oct. 22	654	15.15	9.84	2.72	61.05	
76 ..... {	Oct. 22	664	17.20	11.38	2.59	66.16	Gooseneck from South Carolina; planted May 8. Large canes; the most popular variety in some parts of the South; did not ripen well.
	Oct. 24	681	18.00	11.78	2.66	65.44	
230..... {	Aug. 30	45	14.29	7.65	3.03	53.53	Waubansee. This was formerly a popular variety in Kansas, but has lost favor; large canes, strongly rooted ; large seeds and heavy seed-heads.
	Sept. 17	221	15.59	10.38	1.42	66.58	
	Sept. 21	....	16.07	10.82	1.37	67.33	
	Oct. 5	468	16.32	11.71	.91	71.75	
	Oct. 6	498	15.87	11.10	.92	69.94	
	Oct. 15	....	15.77	10.94	1.21	69.37	
....	Oct. 10	545	20.25	13.80	2.84	68.14	Texas Red, received from Arkansas ; large cane of the Honduras type; seed bright red. Was not grown at the station, and the analysis given was a single cane in a lot received from Arkansas. The high percentage of sugar in connection with the great size of the cane would seem to lend interest to its further investigation.
220	May 19	....	....	....	....	....	White African. Analysis given under experiments in development, page 122.

UNNAMED VARIETIES.

In many cases packages of sorghum-seed were received and planted in the experimental field, which showed as they matured that they had been wrongly named. When the plot was seen to be identical beyond doubt with some other well-known variety, it would be classed with it, as has been done in several cases above. But this could not be decided definitely in all cases. Varieties were also received for identification which were unknown to us.

Ten varieties were received from Algiers, the names of which could not be given by the person who sent them.

Probably the most interesting collection of seed received by the station was furnished by Dr. Peter Collier, director of the New York experiment station. It comprised a large number of varieties, including many from foreign countries, collected through consuls while Dr. Collier was in the U. S. Department of Agriculture. Unfortunately the names and records of these varieties could not be obtained.

Some of these unnamed varieties could perhaps be pretty closely identified, but it is thought better to give them at present just the numbers of the experimental plots where they grew until they can be more certainly identified by another season's planting.

Plot.	Date.	No. of analysis.	Degree Brix.	Sucrose.	Glucose.	Coefficient of purity.	Remarks.
				Per cent.	Per cent.		
9......	Sept. 10	138	14.13	7.40	2.76	52.37	Planted May 8. Canes irregular; badly mixed. From this country.
	Sept. 18	234	15.72	10.58	1.06	67.30	
	Oct. 15	010	16.26	11.48	1.29	70.60	
	Oct. 19	041	14.50	8.01	2.04	55.24	
11......	Sept. 18	237	9.68	3.66	3.04	37.81	Planted May 8. Remarkably large, but short canes, very heavy seed-heads; not a promising variety. From South Africa.
	Oct. 8	513	12.20	6.55	2.65	53.69	
	Oct. 15	611	10.56	3.80	3.34	35.98	
	Oct. 19	640	10.50	4.05	2.80	38.57	
14......	Sept. 11	150	17.67	10.46	3.25	59.20	Planted May 8. Large, handsome canes, free from all offshoots until over-ripe; a promising variety. From this country.
	Sept. 18	245	16.54	9.93	3.75	56.41	
	Sept. 26	346	18.72	12.58	2.48	67.20	
	Oct. 6	590	16.37	10.00	2.92	61.09	
	Oct. 15	606	17.50	11.31	.77	64.63	
	Oct. 19	646	18.60	13.84	.55	74.41	
15......	Sept. 15	239	17.27	11.39	.77	65.95	Planted May 8. Every cane formed several heads; light seed-bearer; notable in its low content of glucose. From India.
	Sept. 26	347	18.64	12.87	.65	69.05	
	Oct. 15	606	17.50	11.31	.77	64.63	
	Oct. 19	642	17.10	11.24	.72	65.73	
16......	Sept. 1	73	13.49	7.57	1.94	56.12	Planted May 8. Same habit as No. 15. From South Africa.
	Oct. 19	644	16.10	10.70	1.31	66.46	
22......	Sept. 1	72	14.84	8.60	2.55	57.95	Planted May 8. Large stocky canes, free from offshoots; heavy seed-top. From this country.
	Sept. 18	235	14.80	9.05	2.16	61.15	
	Oct. 6	496	14.92	10.40	1.74	69.71	
	Oct. 19	643	17.09	11.54	1.49	67.52	
24......	Sept. 1	236	15.70	8.57	3.08	54.59	Planted May 8. Large and fine canes, free from offshoots until over-ripe; light seed-bearer. From Africa.
	Oct. 6	497	18.42	12.72	2.86	69.06	
	Oct. 15	609	17.10	11.09	2.96	64.85	
	Oct. 23	673	15.69	8.53	1.98	51.37	
26......	Sept. 18	238	14.30	6.16	2.24	43.08	Planted May 8. Short, stocky canes; heavy seed-tops. From South Africa.
	Oct. 8	515	16.31	11.48	1.60	70.39	
	Oct. 15	607	13.50	8.14	2.12	60.30	
28......	Oct. 15	608	16.20	11.38	1.41	70.25	Planted May 8. Not a promising variety. From Africa.
33......	Aug. 29	24	12.08	5.77	2.73	47.76	Planted May 8. Good canes, free from offshoots until over-ripe. From this country.
	Sept. 1	66	13.90	7.30	2.80	52.52	
	Sept. 10	137	14.44	7.64	2.80	52.91	
	Sept. 18	211	16.97	10.87	2.08	64.05	
	Sept. 26	351	16.97	11.85	1.41	69.83	
	Oct. 9	521	16.88	11.59	1.21	68.66	
	Oct. 0	522	17.08	11.48	1.81	67.21	
36......	Sept. 1	75	10.14	3.76	1.22	37.08	Planted May 8. The largest canes in the experimental field; did not fully mature. From Africa.
	Oct. 6	494	15.32	10.29	.63	67.17	
	Oct. 19	647	14.06	8.97	.79	63.80	
39......	Sept. 1	76	13.22	6.69	1.54	50.61	Planted May 8. Good canes; few offshoots. From Africa. Remarkable from its low content of glucose and high purity.
	Sept. 18	242	16.47	10.94	1.06	66.42	
	Sept. 26	345	18.32	13.32	.67	72.71	
	Oct. 5	480	17.20	12.79	.60	74.36	
	Oct. 15	603	16.70	12.19	.60	72.99	
44......	Aug. 25	13	14.85	9.39	1.88	63.23	Planted May 8. Good canes, tall and slender; mixed varieties. From this country.
	Sept. 1	77	18.23	12.29	1.23	67.42	
	Sept. 1	78	17.26	11.07	1.80	64.14	
	Sept. 1	79	17.36	10.83	1.36	62.38	
	Sept. 1	80	16.73	11.29	1.42	67.48	
	Sept. 7	121	15.62	10.12	1.44	64.79	
	Sept. 11	149	18.40	12.51	1.25	67.90	
	Sept. 18	249	14.97	9.05	1.86	60.45	
	Sept. 26	356	17.44	11.87	1.05	68.06	
	Oct. 9	527	17.58	11.71	1.22	66.61	
47......	Sept. 18	243	17.52	11.24	.73	64.16	Planted May 8. Similar to No. 15.

Plot.	Date.	No. of analysis.	Degree Brix.	Sucrose.	Glucose.	Coefficient of purity.	Remarks.
				Per cent.	*Per cent.*		
48......	Sept. 1	82	16.71	9.42	3.73	56.37	Planted May 8. Diff. rs but slightly from Orange canes.
	Sept. 11	144	14.50	7.34	3.28	50.62	
	Sept. 18	246	18.45	13.09	2.04	70.93	
	Sept. 26	350	19.21	13.52	1.08	70.38	
	Aug. 29	20	16.68	4.90	2.65	45.88	
	Sept. 1	69	13.82	7.65	2.72	55.35	
50......	Sept. 1	70	12.32	5.07	3.40	41.15	Planted May 8. Mixed canes.
	Sept. 1	71	17.34	10.76	2.90	62.05	
	Sept. 18	252	17.02	12.15	1.25	71.39	
	Oct. 6	495	17.84	12.00	1.66	72.31	
	Oct. 9	524	18.00	13.28	1.01	73.78	
	Sept. 10	141	14.41	7.95	2.90	55.17	Resembles Chinese, but shorter and more stocky canes. From this country.
	Sept. 18	250	17.07	11.26	3.22	65.96	
	Sept. 26	352	17.44	11.73	2.66	67.26	
51......	Oct. 6	403	17.77	12.60	2.27	72.03	
	Oct. 15	599	16.87	11.54	2.25	68.41	
	Oct. 19	632	17.78	12.65	2.27	71.15	
	Oct. 22	665	16.00	11.19	1.99	60.94	
	Sept. 18	233	18.07	11.66	.91	64.53	
	Oct. 6	492	15.47	10.67	.99	68.97	
52......	Oct. 9	526	16.70	11.02	.61	65.99	Similar to No. 15. Planted May 8.
	Oct. 15	507	17.37	11.08	.66	63.79	
	Oct. 19	633	17.50	11.15	.97	63.71	
	Oct. 20	652	16.85	10.47	1.02	62.14	
	Sept. 18	251	13.47	4.90	4.68	36.38	Planted May 8. Strong stocky canes; large seed-heads; good canes.
53......	Oct. 6	490	16.85	10.57	2.96	62.73	
	Oct. 15	598	14.37	7.18	4.00	49.97	
	Oct. 19	635	12.08	4.81	4.01	39.82	
	Sept. 19	260	14.77	6.15	4.72	41.64	Planted May 8. Good canes; light seed-bearer, spangled seed-top; no offshoots until over-ripe. From this country.
	Sept. 26	349	17.70	9.52	4.24	53.79	
57......	Oct. 6	480	18.20	9.58	3.65	52.64	
	Oct. 15	593	18.36	9.97	3.87	54.30	
	Oct. 19	634	18.00	9.93	3.50	55.17	
61......	Oct. 6	487	18.78	13.07	2.26	69.60	Planted May 8. Large, handsome canes, the finest in the experimental field, has the smallest seed-heads and produces less seed than any other; no offshoots until mature, when it shows a tendency to produce secondary seed-heads; did not fully mature before frost. It will perhaps succeed better when fully acclimated. It is a very promising variety.
	Oct. 15	591	18.92	13.13	2.40	69.40	
	Oct. 19	636	18.60	13.06	2.32	70.22	

DEVELOPMENT OF SORGHUM.

Four plots of different varieties were selected for the purpose of making frequent analyses to trace the development of the canes. The analyses were begun the first week in September, and samples were taken every other day until after frost. The results are given in the following tables:

Development of White African, plot 229.

Date.	No. of analysis.	Degree Brix.	Sucrose.	Glucose.	Coefficient of purity.	Remarks.
			Per cent.	*Per cent.*		
Sept. 4	91	14. 64	7. 47	2. 52	51. 02	Seed soft.
Sept. 6	106	15. 30	7. 95	2. 66	51. 96	Do.
Sept. 8	124	14. 78	7. 87	2. 66	53. 24	Do.
Sept. 11	143	14. 70	8. 26	2. 22	56. 20	Seed getting hard.
Sept. 13	168	14. 82	8. 06	2. 24	54. 30	Do.
Sept. 15	194	15. 53	9. 11	2. 02	58. 66	Seed hard.
Sept. 18	230	14. 10	7. 19	2. 32	50. 90	Seed mature.
Sept. 20	276	14. 12	7. 19	2. 42	50. 92	Seed hard.
Sept. 22	305	17. 20	11. 10	1. 67	64. 54	Do.
Sept. 25	327	15. 82	9. 68	1. 18	61. 19	Seed brittle.
Sept. 27	365	14. 27	7. 33	2. 42	51. 36	Do.
Sept. 29	402	13. 58	6. 66	2. 44	49. 04	Do.
Oct. 2	433	12. 87	5. 15	2. 32	40. 01	Do.
Oct. 4	452	12. 23	5. 81	2. 34	47. 50	Do.
Oct. 6	485	11. 67	5. 22	2. 22	44. 73	Do.
Oct. 9	519	13. 39	8. 63	2. 22	64. 45	Do.
Oct. 11	555	14. 18	7. 87	2. 26	55. 50	Do.
Oct. 13	583	15. 76	9. 14	1. 85	58. 00	Do.

Development of Link's Hybrid, plot 0.

Date.	No. of analysis.	Degree Brix.	Sucrose.	Glucose.	Coefficient of purity.	Remarks.
Sept. 3	88	15. 59	8. 53	2. 13	51. 71	Seed soft.
Sept. 5	96	14. 20	7. 78	2. 41	54. 79	Do.
Sept. 7	117	15. 27	9. 37	1. 94	61. 36	Do.
Sept. 10	135	15. 42	9. 28	2. 21	60. 18	Do.
Sept. 12	154	16. 60	10. 18	1. 96	63. 63	Do.
Sept. 14	184	15. 20	10. 22	1. 86	67. 24	Seed getting hard.
Sept. 17	215	15. 75	10. 41	1. 75	66. 10	Do.
Sept. 19	257	16. 15	10. 55	1. 56	65. 33	Do.
Sept. 21	285	17. 35	12. 21	1. 43	70. 37	Do.
Sept. 24	314	18. 00	12. 99	1. 01	72. 17	Seed becoming brittle.
Sept. 26	338	18. 08	12. 75	1. 33	70. 52	Seed brittle.
Sept. 28	388	19. 02	13. 99	1. 11	73. 55	Do.
Oct. 1	410	18. 88	13. 62	1. 14	72. 14	Do.
Oct. 3	413	18. 92	14. 09	. 90	74. 47	Do.
Oct. 5	466	18. 45	13. 97	. 82	75. 72	Do.
Oct. 8	507	18. 38	13. 50	1. 06	73. 45	Do.
Oct. 10	513	18. 05	13. 27	. 79	73. 52	Do.
Oct. 12	571	18. 72	13. 91	. 75	74. 31	Do.
Oct. 15	585	16. 64	11. 49	1. 40	69. 05	Do.
Oct. 17	620	17. 31	12. 24	. 99	70. 71	Do.
Oct. 20	651	18. 26	13. 45	. 50	73. 66	Do.
Oct. 23	671	17. 31	12. 15	1. 29	70. 19	Do.
Oct. 24	678	17. 60	11. 65	1. 60	66. 19	Do.
Oct. 28	688	16. 83	9. 45	.....	56. 15	Do.
Oct. 30	694	20. 40	8. 25	6. 01	40. 44	Do.

Development of Honeydrip, plot 100.

Date.	No. of analysis.	Degree Brix.	Sucrose.	Glucose.	Coefficient of purity.	Remarks.
			Per cent.	*Per cent.*		
Aug. 31	59	11.47	5.14	3.06	44.81	Seed soft.
Aug. 31	60	14.52	6.06	2.64	41.74	Do.
Sept. 4	92	13.50	6.08	3.65	45.04	Do.
Sept. 6	107	15.20	8.05	3.39	52.95	Do.
Sept. 8	131	14.99	8.13	3.10	54.24	Do.
Sept. 11	142	14.87	7.91	3.09	53.19	Do.
Sept. 13	167	15.23	7.91	3.04	52.13	Do.
Sept. 15	196	15.12	8.05	3.32	53.24	Seed getting hard.
Sept. 18	2.2	16.20	9.57	2.50	59.07	Seed mature.
Sept. 20	277	16.62	10.98	2.64	66.06	Seed brittle.
Sept. 22	303	17.85	11.75	2.96	65.83	Do.
Sept. 25	326	15.06	8.48	2.44	56.31	Do.
Sept. 27	364	16.77	10.22	2.30	60.94	Do.
Sept. 29	403	15.82	9.09	2.61	57.46	Do.
Oct. 2	432	16.89	10.40	2.05	63.94	Do.
Oct. 4	453	18.06	11.29	2.12	62.51	Do.
Oct. 6	484	17.74	12.41	1.60	69.95	Do.
Oct. 9	520	17.38	11.37	1.91	65.42	Do.
Oct. 11	556	16.18	9.71	1.87	60.01	Do.
Oct. 13	581	18.56	12.46	1.40	67.13	Do.
Oct. 16	614	17.15	10.74	1.58	62.62	Do.
Oct. 19	629	16.85	10.75	1.58	63.80	Do.
Oct. 22	666	15.70	10.62	1.62	67.64	Do.
Oct. 28	684	16.80	9.11	3.46	54.23	Do.
Oct. 30	690	14.60	5.31	5.24	36.37	Do.
Oct. 30	697	16.80	6.29	4.24	37.44	Do.

Development of Early Orange, plot 223.

Date.	No. of analysis.	Degree Brix.	Sucrose.	Glucose.	Coefficient of purity.	Remarks.
Sept. 7	119	16.36	10.03	3.07	61.31	Seed soft.
Sept. 10	136	15.52	9.17	3.20	59.09	Do.
Sept. 12	155	16.67	10.59	2.71	63.53	Seed getting hard.
Sept. 14	181	17.70	11.90	2.48	67.23	Do.
Sept. 17	217	16.52	10.72	2.63	64.89	Seed mature.
Sept. 19	258	17.97	12.47	2.02	69.39	Do.
Sept. 21	286	17.04	11.01	2.67	64.61	Do.
Sept. 24	315	18.12	12.21	2.48	67.38	Seed brittle.
Sept. 26	339	19.26	13.05	2.32	70.87	Do.
Sept. 28	369	18.52	13.05	2.09	70.46	Do.
Oct. 1	420	19.70	13.67	2.10	69.39	Do.
Oct. 3	444	17.45	12.57	2.09	72.03	Do.
Oct. 5	467	18.76	13.62	1.2	72.60	Do.
Oct. 8	506	18.60	13.25	1.84	71.24	Do.
Oct. 10	544	18.77	13.57	1.48	72.30	Do.
Oct. 12	570	18.82	13.61	1.48	72.32	Do.
Oct. 15	586	18.64	13.08	1.97	70.17	Do.
Oct. 17	619	18.62	13.08	1.60	70.25	Do.
Oct. 19	628	18.29	13.15	1.75	71.90	Do.
Oct. 22	662	16.50	11.11	2.17	67.31	Do.
Oct. 23	672	17.35	11.56	1.81	66.63	Do.
Oct. 24	679	17.60	10.60	2.91	60.23	Do.
Oct. 28	685	16.60	6.73	4.80	40.54	Do.
Oct. 30	693	16.30	8.49	3.50	52.09	Do.
Oct. 30	695	15.50	6.17	4.48	39.81	Do.

Considerable work has already been published on the subject of the development of sorghum cane and the changes undergone by the different constituents of the juice as the cane approaches maturity. The great difficulty in all such work comes from the liability to error in sampling. A sample taken one day may show a lower content of sucrose than one taken the day before, not because the cane has all undergone deterioration, but because the sample taken on the second day was composed of poorer canes than the first. Moreover, there is undoubtedly some change brought about in the quality of the juice by conditions of the soil and atmosphere, causing a difference in the amount of water contained by the plant at different times. It is quite an important matter, however, to ascertain the point in the development of the plant when it contains its maximum of sugar, especially if this point is coincident with some outward appearance of the plant by which the proper time for working it up may be known; and the difficulties in the way should not deter us from a progress in that direction. It has been asserted that the sorghum cane is ready for working as soon as the seed is mature. Our work on development, and observations in general would lead us to doubt the truth of this tenet. The juice does not attain its maximum of sucrose or of purity until long after the seed is sufficiently mature to germinate. From the above tables it will be seen that the canes improved materially after the seed had become perfectly hard and brittle, and after the appearance of the canes in general would have led most practical sorghum men to pronounce them overripe. With the exception of the White African, none of the varieties used commenced to deteriorate until after there had been heavy frosts, about the middle of October. All were varieties that are rather late in maturing, as work was not commenced in time to follow the development of any of the earlier varieties. There is little doubt but that the later varieties are generally harvested too early in the work of the factories; and the necessity is evident of either making late successive plantings of the earlier varieties, planting varieties which ripen intermediately between the early and late, or of selecting from the later varieties with a view to their earlier maturation. It is a pretty general observation in our experience that the ripeness of sorghum cane is overjudged when based upon its external appearance, and doubtless many of the published analyses which have brought disrepute to sorghum as a sugar-producing plant, aside from the cases where it was grown too far north to permit of its maturing, were made upon canes which had been cut when the seed, but not the juice, was mature.

SUMMARY OF THE ANALYSES OF DIFFERENT VARIETIES.

In the following table the highest result attained by average samples from plots of the different varieties grown is given. In nearly all cases the sample showing the highest content of sugar gave also the best results in the other two essentials, viz, minimum of glucose and maximum of purity; but where this rule did not hold good, the analysis which showed superiority in two essentials was inserted as the maximum analysis attained by the variety during the season.

Maximum analysis of each variety.

Variety.	No. of plot.	Date.	No. of analy-sis.	Degree Brix.	Su-crose.	Glu-cose.	Coeffi-cient of purity.
					Pr.cent.	*Pr.cent.*	
Swain's Early Golden	93	Aug. 24	3	18 03	12.88	1.92	71.44
Early Tennessee	102	Aug. 24	6	15.54	8.45	2.55	51.38
Whiting's Early variety	234	Aug. 30	46	15.63	10.30	1.48	65.90
Black Amber	90	Sept. 7	120	12.88	6.50	3.14	50.47
White Amber	92	Aug. 24		18.03	12.40	1.84	72.05
Early Amber from New York	23	Aug. 24	4	18.10	13.70	1.12	75.69
Early Amber from Kansas	1	Aug. 25	8	17.54	13.18	1.07	75.14
Folger's Early variety	249	Oct. 4	460	16.78	11.01	1.46	71.16
Chinese from Central America	63	Oct. 15	593	17.37	11.79	1.38	67.88
Chinese from New South Wales	215	Sept. 17	216	16.29	9.76	2.25	59.91
Chinese from Africa	232	Oct. 8	508	17.98	12.46	1.41	69.20
Chinese from United States	37	Oct. 8	511	19.00	13.23	1.49	69.63
White India	69	Oct. 9	531	17.67	13.07	1.02	73.96
White India from Louisiana	67	Oct. 5	474	16.33	11.90	1.20	72.87
Early Orange from Kansas	84	Sept. 27	371	17.58	12.82	1.33	72.94
Early Orange from South Carolina	228	Oct. 5	467	18.76	13.62	1.72	72.60
Early Orange from Arkansas	87	Sept. 19	270	16.53	11.30	2.20	68.91
Early Orange from Louisiana	68	Sept. 15	601	17.00	12.90	1.15	72.07
Kansas Orange	49	Sept. 9	523	16.91	12.17	1.20	71.97
New Orange	88	Sept. 27	375	16.25	9.53	3.07	58.65
Late Orange from New Jersey	89	Oct. 9	540	17.90	12.73	2.32	70.76
Medium Orange	235	Sept. 26	341	16.70	11.84	1.04	70.90
Red Liberian from Missouri	72	Oct. 4	462	18.80	13.25	2.74	70.48
Red Liberian from Texas	73	Sept. 27	367	19.92	14.76	1.84	74.10
Golden Rod	95	Sept. 5	99	14.35	6.96	4.48	48.50
Honey Dew	250	Oct. 22	660	18.10	11.92	2.02	65.86
Dutcher's Hybrid	98	Sept. 3	85	16.24	10.50	2.14	61.66
Link's Hybrid	0	Oct. 5	466	18.45	13.97	.82	75.72
Price's Hybrid	101	Sept. 3	84	16.84	10.77	2.71	63.95
Planter's Friend	214	Sept. 30	418	20.50	13.84	1.66	67.46
Honduras from Louisiana	64	Oct. 15	595	15.54	9.54	3.24	61.30
Honduras from Texas	66	Oct. 22	654	15.15	9.84	2.72	64.95
Gooseneck	76	Oct. 22	664	17.20	11.38	2.59	66.16
Waubansee	230	Oct. 5	468	16.32	11.71	.91	71.75
White African	229	Sept. 22	305	17.20	11.10	1.67	64.54
Texas Red		Oct. 10	545	20.25	13.80	2.84	68.14

Unnamed varieties.

Plot No. 9, United States		Oct. 15	610	16.26	11.48	1.29	70.60
Plot No. 11, South Africa		Oct. 8	513	12.10	6.55	2.65	53.69
Plot No. 14, United States		Oct. 19	646	18.00	13.84	.55	74.41
Plot No. 15, India		Sept. 26	347	18.64	12.87	.65	69.05
Plot No. 16, South Africa		Oct. 19	614	16.10	10.70	1.31	66.46
Plot No. 22, United States		Oct. 19	643	17.09	11.51	1.49	67.52
Plot No. 24, Africa		Oct. 6	497	18.42	12.72	2.86	69.00
Plot No. 26, Africa		Oct. 8	515	16 31	11.48	1.60	70.39
Plot No. 28, Africa		Oct. 15	608	16.20	11.88	1.41	70.25
Plot No. 33, United States		Sept. 26	351	16.07	11.85	1.41	69.83
Plot No. 36, Africa		Oct. 6	494	15.32	10.20	.63	67.17
Plot No. 39, Africa		Oct. 5	480	17.20	12.79	.60	74.36
Plot No. 44, United States		Sept. 26	356	17.44	11.87	1.05	68.00
Plot No. 50		Oct. 9	524	18.00	13.28	1.01	73.78
Plot No. 51		Oct. 6	493	17.77	12.80	2.27	72.03
Plot No. 53		Oct. 6	490	16.85	10.57	2.90	62.73
Plot No. 57, United States		Oct. 19	634	18.00	9.93	3.50	55.17
Plot No. 61		Oct. 19	636	18.00	13.06	2.32	70.22
Average				16.79	11.60	1.85	69.62

These results are quite interesting as furnishing a means of comparison of the relative merits of the different varieties. The ten varieties which stand highest in each of the three essentials are given below, in the order of their value.

List of ten varieties giving best results.

Variety.	Sucrose.	Variety.	Glucose.	Variety.	Coefficient of purity.
	Per cent.		*Per cent.*		
1. Red Liberian ...	14.76	1. Plot No.14, United States.	.55	1. Link's Hybrid...	75.73
2. Link's Hybrid...	13.97			2. Early Amber....	75.69
3. Plot No. 14	13.84	2. Plot No. 39, Africa	.60	3. Plot No. 14	74.41
4. Planter's Friend	13.83	3. Plot No. 36, Africa	.63	4. Plot No. 39	74.36
5. Texas Red.......	13.80	4. Plot No. 15, India.	.65	5. Red Liberian...	74.10
6. Early Amber....	13.70	5. Link's Hybrid ...	.82	6. White India....	73.96
7. Early Orange....	13.62	6. Wanbanseo.....	.91	7. Plot No. 50	73.78
8. Plot No. 50......	13.28	7. Plot No. 50	1.01	8. Early Orange ...	72.93
9. Chinese	13.23	8. White India ...	1.02	9. Plot No. 51	72.53
10. White India.....	13.07	9. Medium Orange	1.04	10. Kansas Orange..	71.97
		10. Plat No.44, United States.	1.05		

These lists comprehend altogether eighteen varieties, of which four appear in all three of the lists, four on two, and ten on only one, as follows:

Variety.	No.	Variety.	No.	Variety.	No.
Plot No. 14	3	Red Liberian.......	2	Plot No. 44	1
Link's Hybrid	3	Early Orange	2	Plot No. 57	1
Plot 50	3	Plot No. 36	1	Kansas Orange.....	1
White India.........	3	Plot No. 15........	1	Planter's Friend ...	1
Plat No. 39	2	Wanbanseo	1	Texas Red	1
Early Amber.......	2	Medium Orange....	1	Chinese.	1

From this it will be seen that four varieties combine in a high degree the three good qualities of a large percentage of sucrose, low content of glucose, and high purity of juice. Link's Hybrid and the unnamed variety No. 14 divide honors for the first place, both standing very near the top of the list in all three essentials. The former has always proved a good sugar producer where it has had time to mature before frost. The Early Amber is noticeable for its high purity, five of the plats of its subvarieties giving a purity of over 70. From this quality doubtless arises its superiority as a sirup-making variety. The low content of glucose in several of the unnamed varieties from tropical countries is remarkable, as most of them were not entirely mature before frost. It must not be lost sight of in comparing the varieties on the basis of the analyses that the outward faults of a variety may entirely overbalance its value as shown by analysis. The Link's Hybrid, for instance, which gives such good results on analysis, has a fault of form that almost destroys its practical value. This point will be considered further on.

II. EXPERIMENTS IN HYBRIDIZING OR CROSSING VARIETIES.—III.
EXPERIMENTS IN PRESERVING SPORTS OR VARIATIONS.

These two methods of improvement may as well be considered together, for in the present condition of the sorghum plant it is hard to draw the line between them. The different varieties which have become established cross so readily with one another that where variations occur, in a field of cane for instance, it is often difficult to say positively whether it is a true sport, whether it is from one seed of a distinct variety accidentally introduced, or whether it is from a seed that had been cross-fertilized from a different variety. Doubtless both causes of variation obtain to a large extent, for the one is a natural consequence of the other; that is, on account of the readiness with which two individuals cross, a large number of varieties have been produced, and as many of these are not well established or fixed they exhibit a constant tendency to revert to original types, thus showing variations. Whether the wide variations shown in the different kinds of sorghum are due more to crossing or more to type variation, is a question it is unnecessary to discuss here. It is sufficient to show that such capability for variation does exist. In the work done at this station no distinction could be made between variations produced by crossing and those which were true sports. As this season's work was only the beginning it was impossible to obtain true artificially-produced crosses; that is, variations produced by the careful cross-fertilization of two distinct and definite types. The plots called "crosses" were planted from seed-heads obtained by Mr. Denton from various fields of sorghum, and were simply variations from the general type of the cane growing about them. In the great majority of cases the canes produced from this seed showed such well-marked reversions to two well-defined types that it was a pretty fair presumption that they actually did result from the cross-fertilization of those types. But of course such work should, in the future, be carried out upon known types artificially cross fertilized.

GENERAL OBSERVATIONS ON CROSSES.

Kolreuter says, "He who would produce new varieties should cross varieties."

Darwin says: "In regard to the beneficial effect of crosses between varieties there is plenty of evidence." "The crossing of two forms which have long been cultivated implies that new characters actually arise, some of which may be valuable and permanent." "It would be superfluous to quote more, for Gartner, Herbert, Sageret, Lecoq, Naudin, and many other eminent experimenters speak of the wonderful vigor, size, tenacity of life, precocity, and hardiness of hybrid productions."

It is stated in the Sugar Beet* that "if a superior variety of beets be placed near another variety, the result will be most advantageous, and

---

* The Sugar Beet, by Lewis E. Ware.

it may be concluded from these experiments, which we can indorse, that the resulting race will, for the time being, be richer in seed, and that the roots grown therefrom will contain a sugar content, more regular, etc., than had existed in either."

In regard to the effect of crossing varieties, it can be said that it seems to increase the vigor of the plants sometimes in a wonderful degree. The crossed canes are often much larger and taller and often have much heavier seed-heads than either parent form. A crossed cane is sometimes earlier, often later, in maturing than either parent. Some crosses breed true to the new type from the start, and show no tendency to reversion, but usually the first season the crossed seeds are planted some of the plants revert, some to one parent form, some to the other; some are intermediate forms. If, now, seed of the type preferred is selected and planted again, the new plants show less tendency to revert; by continuing the selection and throwing out varying forms the new type is fixed and becomes a new variety. There is greater tendency to reversion in "violent" crosses between dissimilar forms than in crosses of allied forms. A cross may be slight or complete; in fact there may be several crosses between two varieties. For instance, a fixed cross between the Early Amber and the Orange may resemble the Early Amber more. Another cross between the same varieties may resemble the Orange more. Three canes taken from a plot of this last cross showed by analysis a higher percentage of sugar than any other in the season's work, with one exception.

ADVANTAGES OF SORGHUM OVER SUGAR-CANE ON ACCOUNT OF THE EASE WITH WHICH VARIATIONS ARE PRODUCED IN THE FORMER.

Dr. Morris, formerly director of the Jamaica Botanical Gardens, where an experimental plantation of sixty to seventy varieties of the sugar cane is maintained, in an address before the London Chamber of Commerce said:

"It is well known that the sugar cane does not produce seed, and hence it is impossible to improve it by any processes of hybridizing and crossing found so beneficial to other plants. New varieties amongst sugar canes arise generally in the form of bud variation. These occur very seldom, and possibly amongst thousands of acres not one cane will be detected which exhibits any well-marked characteristics. Planters, however, should be keen to notice any canes that show a departure from the types, and should cultivate them separately. If the sugar cane were capable of being improved purely by cultivation and experimental processes like those which have improved the beet, this would be one of the most effective means of benefiting the industry."

GENERAL OBSERVATIONS ON SPORTS OR SPONTANEOUS VARIATIONS.

It is well known that new varieties sometimes suddenly and spontaneously appear in plants. They are created by bud variation. A peach tree suddenly produces a branch which yields nectarines; a plum tree which had yielded yellow plums for forty years produced a single bud which produced a new and valuable permanent variety, the Red Magnum Bonum plum.

The variations in the tropical sugar cane are entirely produced in that way, as has already been shown by the statements of Professor Morris just quoted.

In Mauritius a sugar cane of the ribbon variety produced two new canes, a green cane and a red cane. This was considered an astonishing variation there. The causes of such variations are unknown. It is only known that they do occur, and that valuable new varieties sometimes suddenly appear in that way.

The history of some of the varieties of sorghum would seem to indicate, so far as it is possible to obtain accurate information of such matters, that they originated in this way.

In Indiana, in a field of Chinese cane, a single cane ripened two weeks earlier than the other canes. This variation was preserved and named the Early Amber. It is the most widely known of all the varieties of sorghum. In the experimental field of this station there were growing Early Amber canes received from New South Wales, from Cape Town, and from many places, showing its wide distribution.

In New York, in a field of Early Amber, only one cane ripened before frost. This variation was preserved and named by us Whiting's Early Variety. It matures ten days earlier than the Early Amber. It seems to be a sport from a sport.

In Tennessee, in a field of Honduras, a single cane ripened two weeks earlier than the other canes. This variation was preserved and was named Link's Hybrid. It is one of the best varieties of sorghum for sugar manufacture.

It is probable that other cane-growers have seen as valuable variations in their cane fields, and have not recognized their importance. It is worthy of remark that each of these variations was noticed and was preserved merely because it chanced to ripen earlier than the other canes in the same field, and not because its other qualities were recognized at the time.

. In the effort to improve the sorghum plant all such variations from type should be analyzed to determine their value in sugar manufacture.

WORK AT THE STERLING STATION ON CROSSES OR VARIATIONS.

It may be said of the work done here in this direction that in the first place it established positively, in the judgment of those in charge, *the fact of the very strong tendency of this plant toward variability.* This fact has, of course, been frequently noticed and commented upon heretofore, but as it seems essential that it should be thoroughly and generally understood, we think it advisable to enter into an exposition of the evidence that was obtained to justify us in coming to the very decided conclusion we adopted upon this point. The plots which were planted as " crosses " at this station were in every case from single seed-heads, selected by Mr. Denton, which were very carefully thrashed and cleaned, special precautions being taken to prevent any accidental admixture of seed from other sources. These plots were then in every case the product of a *single head;* they showed, in the majority of cases, the greatest variation among the individual canes.

This variability is well shown by a series of photographs taken by us, which were intended to be reproduced as illustrations of this report. Unfortunately the fund provided for such illustrations was exhausted at the time this bulletin was sent to the press, so that they had to be omitted. They represent a number of seed-heads, all taken from the same plot, which showed striking variations from either parent type, as well as gradations running back to each. In a plot planted from a single seed-head which was evidently a cross between the Orange and India, for instance, heads were selected which gave the greatest variations and gradations between the India type, with its white seeds and rather loose head, to the Orange, with its reddish-colored seeds and compact head. Another represents the range of variations between the Honduras and Red Liberian, two widely different varieties, with the small round seed of the Liberian type set closely on the sprangle top head of the Honduras. These photographs of the widely different types produced from a single seed-head would convince any one, we think, of the great ease with which variations can be produced in sorghum.

LIST OF CROSSES.

The following list gives the number of the experimental plot with the probable parents of some of the crosses grown this season. Many plots are not included, as the characters shown by the canes did not distinctly indicate the origin of the variation.

No. of Plot.	Probable cross.	No. of Plot.	Probable cross.
110	New Orange and Early Orange.	163	India and Orange.
111	Chinese and Liberian.	165	India and Amber.
112	Kansas Orange and Amber.	166	Do.
114	Golden Rod—cross.	167	India—cross.
115	Orange and Amber.	168	Do.
117	Kansas Orange and Amber.	171	Kansas Orange and India.
118	Liberian and Golden Rod.	172	New Orange—cross.
120	Amber and Kansas Orange.	173	India—cross.
121	Orange and White India.	174	India and Amber.
122	Orange and Chinese.	175	New Orange and Early Orange.
124	India—cross.	176	Orange—cross.
127	India and Golden Rod.	178	India and Orange.
128	Do.	179	India—cross.
129	Do.	180	Orange and India.
131	Orange and India.	181	Do.
132	India and Golden Rod.	182	India and Amber.
133	Kansas Orange and India.	183	India—cross.
134	Orange and Golden Rod.	184	Orange and India.
135	Early Orange and Amber.	185	Orange—cross.
136	Orange and India.	186	Orange and India.
137	India and Amber.	187	Do.
138	Do.	188	Do.
139	Orange and India.	193	Orange—cross.
140	Do.	194	Do.
142	India—cross.	195	Orange and India.
144	Orange and Amber.	196	Kansas Orange and India.
146	Kansas Orange and Golden Rod.	197	India—cross.
147	Kansas Orange and New Orange.	200	New Orange—cross.
151	Orange and India.	201	Do.
153	Kansas Orange and Early Amber.	202	India—cross.
154	Amber and New Orange.	204	India and Orange.
155	Orange—cross.	205	Orange—cross.
157	Amber—cross.	208	India—cross
158	India and Orange.	211	Orange and India.
161	Kansas Orange and India.	212	Do.
162	India and Orange.		

ANALYSES OF THE CROSSES.

The following table gives the analyses made of average samples taken from the different plots:

Analyses of Hybrids and Crosses.

No of plot.	Date.	No of analysis.	Degree Brix.	Sucrose.	Glucose.	Co-effi-cient of purity.
				Per cent.	*Per cent.*	
	Aug. 29	31	14.65	9.45	2.19	64.50
	Aug. 31	55	15.16	9.24	2.69	60.95
110	Sept. 6	110	16.42	10.65	2.46	64.86
	Sept. 20	280	18.32	13.08	1.28	71.39
	Sept. 27	374	17.56	12.30	1.48	70.41
	Oct. 9	541	18.04	12.54	1.96	69.51
112	Aug. 29	32	13.82	9.04	1.36	65.41
113	Sept. 12	162	17.42	11.40	2.36	65.44
	Sept. 27	376	18.78	13.49	1.77	71.83
	Sept. 12	163	13.70	6.08	3.76	44.37
115	Sept. 12	164	16.36	10.08	2.57	61.61
	Sept. 24	324	14.22	7.25	3.57	50.98
117	Sept. 27	377	14.26	7.78	3.00	54.56
120	Aug. 31	57	11.50	4.44	3.40	38.60
	Sept. 12	166	13.51	7.63	3.08	56.47
122	Aug. 30	35	8.14	2.42	2.53	29.73
132	Oct. 11	562	15.67	10.90	1.47	69.56
133	Oct. 11	561	16.15	10.39	2.83	64.33
134	Sept. 13	169	13.90	6.96	3.47	63.03
	Oct. 11	560	15.67	9.82	2.18	62.66
135	Oct. 11	559	16.65	11.05	1.96	66.36
136	Oct. 11	558	18.15	12.24	1.82	67.43
137	Oct. 11	557	19.35	14.49	.90	74.88
144	Sept. 27	379	16.06	10.02	2.29	62.39
146	Aug. 30	34	10.72	3.89	3.65	36.29
147	Sept. 12	165	14.70	7.85	3.15	53.40
	Sept. 27	378	16.82	10.97	2.48	65.22
148	Sept. 6	112	13.09	6.34	2.30	48.43
152	Aug. 31	52	12.63	6.30	2.38	49.92
	Sept. 6	109	13.34	6.93	1.96	51.95
153	Sept. 13	172	15.97	10.38	1.35	64.99
	Sept. 13	179	15.63	9.58	2.12	61.29
154	Aug. 30	36	14.25	7.49	3.35	52.56
	Sept. 6	108	14.63	8.65	2.95	59.12
	Sept. 13	171	15.80	9.65	2.23	61.07
205	Sept. 15	201	16.33	9.77	3.41	59.83
	Sept. 20	407	16.94	10.77	2.34	63.58
206	Sept. 15	200	12.73	6.20	3.12	48.70
207	Sept. 15	203	11.52	4.84	3.01	42.01
208	Sept. 28	18	12.60	6.06	2.59	48.10
	Sept. 15	205	17.23	11.35	1.53	65.87
209	Sept. 15	206	13.92	8.24	2.37	59.20
210	Sept. 15	210	16.30	8.35	1.90	54.57
211	Sept. 15	209	14.40	7.74	2.80	53.76
	Sept. 20	281	14.82	8.75	2.39	59.04
212	Sept. 15	211	14.53	8.67	1.90	59.67
158	Sept. 14	180	15.62	9.85	2.69	63.06
161	Sept. 13	173	13.06	6.74	1.83	51.61
	Sept. 14	183	15.72	9.80	1.68	62.34
162	Sept. 14	186	15.90	9.04	1.84	62.51
163	Sept. 14	185	15.90	10.18	2.16	64.02
165	Aug. 30	38	10.75	5.09	1.53	47.35
	Sept. 13	170	15.60	9.95	.88	63.78
	Sept. 20	404	16.70	10.92	.90	65.39
166	Aug. 30	39	10.32	3.90	2.65	37.70
	Sept. 13	174	12.04	6.27	1.98	52.08
168	Aug. 28	22	12.67	5.91	3.00	46.64
	Sept. 13	175	14.92	9.94	1.04	66.62
170	Aug. 30	37	11.38	5.18	3.30	45.52
	Sept. 13	176	13.47	8.37	1.68	62.14
	Sept. 15	204	13.82	8.30	1.74	60.05
171	Sept. 13	177	15.07	7.93	3.20	52.55
	Sept. 14	188	14.20	7.29	3.13	51.34
172	Sept. 13	178	14.04	7.46	3.61	53.13
	Sept. 14	182	14.82	8.31	3.08	56.07
173	Aug. 31	53	11.32	4.24	3.64	37.45
	Aug. 31	51	12.76	5.80	4.42	45.45
185	Sept. 14	187	14.40	7.63	2.06	52.98
	Sept. 20	279	15.70	9.06	2.21	57.70
191	Sept. 14	189	14.40	7.49	3.02	52.01
193	Sept. 14	190	12.73	5.90	2.95	46.34

*Analyses of Hybrids and Crosses—*Continued.

No. of plot.	Date.	No. of analysis.	Degree Brix.	Sucrose.	Glucose.	Co-efficient of purity.
				Per cent.	*Per cent.*	
194	Sept. 14	191	12. 93	6. 54	1. 33	50. 58
196	Sept. 15	198	16. 82	11. 12	2. 34	66. 11
197	Sept. 15	197	14. 62	7. 53	3. 64	51. 50
199	Sept. 15	199	16. 12	9. 83	2. 72	60. 98
200	Sept. 15	202	13. 00	6. 05	3. 15	46. 54
202 {	Aug. 28	21	13. 04	6. 82	3. 55	52. 30
	Sept. 15	207	15. 54	9. 42	2. 65	60. 61
203 {	Aug. 30	40	12. 64	6. 35	2. 63	50. 23
	Sept. 15	208	13. 42	7. 57	1. 71	56. 41

After a number of these analyses of large samples had been made, it was concluded to discontinue them, for the individual canes varied so much that it was impossible to obtain samples which would represent the plot, except in the few cases where the character of the plot was uniform. In some cases the plot could be thrown out, where the average samples showed a very poor analysis. The work was thenceforth confined to analyses of individual canes, selected with a view to permanence of type. A very large number of samples were taken in this way, the seed heads removed, marked with a number corresponding to the analysis and preserved. The juice was polarized, and from each plot one or more samples which gave the best results, and which were to be reserved for future planting, were subjected to complete analysis, so as to have a complete pedigree of the cane. The following table gives the results of some of the individual canes from the crosses; only the best samples in each plot are given, and these analyses are only a fraction of the whole number made and recorded at the station:

Analyses of Crosses.

No. of plot.	Date.	No. of analysis.	Degree Brix.	Sucrose.	Glucose.	Co-efficient of purity.
				Per cent.	*Per cent.*	
109	Sept. 24	463	16. 54	10. 31	2. 31	62. 33
	Oct. 10	1069	18. 50	13. 11	....	70. 86
113 {	Oct. 10	1070	18. 00	13. 24	1. 16	73. 56
	Oct. 10	1073	18. 50	13. 16	....	71. 14
120 {	Sept. 28	535	17. 12	11. 30	....	66. 60
	Sept. 28	537	15. 12	9. 43	....	62. 37
	Sept. 24	469	17. 00	11. 37	1. 12	66. 88
123 {	Sept. 28	539	18. 12	12. 61	....	69. 59
	Sept. 28	531	17. 20	11. 33	1. 73	65. 87
	Sept. 28	542	18. 78	13. 63	....	72. 58
124	Sept. 28	546	18. 28	13. 43	....	73. 47
	Sept. 28	549	16. 41	11. 33	....	69. 04
128 {	Oct. 10	1049	17. 30	12. 95	....	74. 86
	Oct. 10	1050	18. 50	13. 81	1. 20	74. 65
	Oct. 10	1052	17. 35	12. 63	....	72. 80
129 {	Sept. 28	550	13. 81	6. 87	....	49. 75
	Sept. 28	552	15. 66	9. 45	....	60. 34
	Oct. 10	1046	17. 00	12. 65	....	74. 41
	Sept. 28	558	18. 37	13. 37	....	72. 78
130 {	Oct. 10	1040	18. 88	14. 11	....	74. 74
	Oct. 10	1041	18. 78	13. 96	....	74. 33
	Oct. 10	1042	20. 48	15. 20	1. 09	74. 21
	Sept. 28	560	19. 37	14. 29	....	73. 77
	Sept. 28	561	19. 37	14. 39	1. 18	74. 29
131 {	Oct. 10	1054	19. 00	13. 66	....	71. 89
	Oct. 10	1055	19. 60	14. 39	....	73. 42
	Oct. 10	1058	19. 20	13. 65	....	71. 09
	Oct. 10	1059	20. 00	14. 48	2. 03	72. 40

133

Analyses of Crosses—Continued.

No. of plot.	Date.	No. of analysis.	Degree Brix.	Sucrose.	Glucose.	Co-efficient of purity.
				Per cent.	*Per cent.*	
132 {	Sept. 28	562	19.42	13.52	1.65	60.62
	Sept. 28	563	18.37	12.90	....	70.22
133 {	Sept. 28	565	15.55	8.87	....	57.04
	Oct. 11	1220	20.65	15.03	1.03	72.78
135 {	Sept. 29	571	16.07	10.94	....	08.08
	Sept. 29	572	19.57	13.30	....	08.42
	Sept. 29	573	19.87	14.34	....	72.17
	Sept. 29	574	20.37	14.01	1.29	69.21
	Sept. 29	577	18.40	12.58	....	68.37
	Oct. 11	1218	21.13	16.33	.77	77.28
136 {	Sept. 29	578	17.00	12.51	....	09.89
	Sept. 29	582	16.00	10.50	....	67.50
	Sept. 29	583	17.00	12.36	....	72.71
	Oct. 11	1200	19.03	13.00	....	68.79
137 {	Sept. 29	584	18.70	13.08	....	73.16
	Sept. 29	585	20.20	15.32	1.21	75.84
	Sept. 29	586	21.50	16.20	.81	75.63
	Sept. 29	587	18.70	13.50	....	72.67
	Sept. 29	588	18.00	13.24	....	73.56
	Sept. 29	589	18.68	13.54	....	72.48
138 }	Sept. 29	590	17.78	13.45	.09	75.60
	Sept. 29	593	17.18	11.44	....	66.59
139	Oct. 13	1043	21.21	13.75	....	64.83
	Sept. 29	612	19.88	14.41	....	72.64
	Sept. 29	613	20.78	15.54	2.33	74.78
142 {	Sept. 29	614	19.14	13.62	....	71.16
	Sept. 29	616	19.81	14.31	....	72.24
	Oct. 10	1104	20.07	14.95	1.08	74.49
	Oct. 10	1103	18.07	13.14	....	72.71
143	Oct. 10	1092	19.00	13.86	....	72.95
144 {	Sept. 29	621	19.28	14.08	1.10	70.14
	Sept. 29	623	17.38	11.92	....	68.58
	Sept. 29	624	16.98	11.54	....	67.96
145 {	Oct. 10	1095	18.20	12.74	....	70.00
	Oct. 10	1099	18.27	13.71	....	75.01
146 {	Sept. 29	625	19.28	13.91	1.82	72.15
	Sept. 29	628	18.70	13.07	....	69.89
	Oct. 10	1087	17.60	12.39	....	70.40
147 {	Sept. 29	602	16.58	11.47	....	69.18
	Sept. 29	603	15.87	11.61	1.60	73.16
	Sept. 29	606	18.28	12.41	....	67.89
14- {	Sept. 29	607	17.08	12.58	....	71.15
	Sept. 29	611	20.88	15.78	1.63	75.57
	Oct. 1	640	18.20	11.29	....	62.03
151 {	Oct. 11	1233	20.30	14.75	....	72.66
	Oct. 11	1237	19.20	13.60	....	70.83
	Oct. 11	1242	21.33	14.75	1.57	69.15
152	Oct. 1	647	18.00	12.36	....	68.67
153 {	Oct. 1	651	18.00	13.33	....	74.06
	Oct. 1	652	22.50	17.18	.58	76.36
	Oct. 11	1134	22.50	16.85	.01	74.89
155 {	Oct. 1	660	20.09	14.27	1.06	71.03
	Oct. 1	661	16.82	11.90	....	70.75
150 {	Oct. 1	663	16.63	10.00	....	60.13
	Oct. 1	664	18.62	12.15	4.50	65.25
157 {	Oct. 1	666	18.29	10.00	....	59.60
159 {	Oct. 1	673	17.65	12.49	1.45	70.70
	Oct. 1	674	18.17	12.41	....	68.30
161 {	Oct. 11	1277	20.70	15.40	.80	74.40
	Oct. 11	1287	19.70	14.89	.81	75.58
162 {	Oct. 1	691	18.85	13.84	1.24	73.42
	Oct. 1	695	16.90	12.00	....	71.01
103 {	Oct. 1	705	17.97	12.07	....	72.18
	Oct. 1	706	20.47	15.53	1.30	75.87
	Oct. 1	707	18.40	13.09	....	71.14
165 {	Oct. 1	710	19.37	14.29	....	73.77
	Oct. 1	711	19.47	14.52	.53	74.58
	Oct. 1	714	18.50	14.06	....	70.00
	Oct. 12	1332	19.42	14.35	.60	73.80
160 {	Oct. 1	719	18.50	12.05	....	70.00
	Oct. 1	722	16.50	11.50	....	70.24
	Oct. 1	724	18.00	12.78	....	71.00
	Oct. 1	725	18.03	12.83	.. ....	71.16
	Oct. 1	726	16.52	13.40	....	81.11
	Oct. 13	1480	20.00	14.15	....	70.75
	Oct. 13	1483	20.50	14.80	....	72.20
	Oct. 13	1491	19.78	15.04	.75	76.04

Analyses of Crosses—Continued.

No. of plot.	Date.	No. of analysis.	Degree Brix.	Sucrose.	Glucose.	Co-efficient of purity.
				Per cent.	Per cent.	
167	Oct. 1	730	17.98	12.55	....	60.80
	Oct. 1	731	17.96	11.77	....	65.53
	Oct. 1	735	16.68	12.08	....	72.42
168	Oct. 1	737	18.20	13.51	....	74.23
	Oct. 1	738	18.60	14.98	.75	80.54
171	Oct. 2	748	18.00	12.57	....	69.83
	Oct. 2	749	18.82	13.06	....	74.18
	Oct. 2	750	16.93	12.13	....	71.65
	Oct. 2	751	17.33	12.00	....	69.24
173	Oct. 2	753	20.30	14.44	1.53	71.13
	Oct. 2	754	19.00	13.51	....	71.11
	Oct. 2	758	19.83	13.89	....	70.05
	Oct. 10	1067	21.20	14.28	1.00	67.36
	Oct. 12	1429	23.00	14.77	....	64.22
	Oct. 12	1430	22.00	15.73	....	71.50
	Oct. 12	1431	22.50	16.28	....	72.36
	Oct. 12	1433	21.70	16.29	1.19	75.07
174	Oct. 2	762	18.43	13.25	2.02	71.89
	Oct. 2	763	18.03	12.35	....	68.50
	Oct. 12	1412	21.60	14.86	2.05	68.80
	Oct. 12	1417	20.78	14.32	....	68.91
175	Oct. 2	769	18.20	12.11	2.82	66.54
	Oct. 2	771	17.50	11.61	....	66.34
176	Oct. 2	776	19.93	14.81	1.76	74.31
	Oct. 2	777	18.35	12.97	....	70.68
	Oct. 2	778	18.05	14.14	....	74.62
	Oct. 2	784	19.00	13.91	....	73.21
	Oct. 2	788	19.00	13.33	....	70.16
178	Oct. 2	789	19.50	14.32	1.29	73.44
	Oct. 2	793	18.70	13.54	....	72.41
	Oct. 2	795	17.20	12.41	....	72.15
	Oct. 13	1560	21.00	15.81	....	75.29
179	Oct. 2	802	17.20	11.97	....	69.59
	Oct. 2	804	17.20	12.31	....	71.57
180	Oct. 2	811	20.00	14.14	1.05	70.70
	Oct. 2	812	18.50	13.26	....	71.68
181	Oct. 2	818	17.00	11.78	....	69.29
	Oct. 2	820	17.00	10.47	....	61.59
182	Oct. 12	1357	20.28	13.96	1.53	68.84
184	Oct. 12	1339	20.45	14.97	....	73.20
	Oct. 12	1340	22.26	15.70	....	70.53
	Oct. 12	1342	20.96	14.90	....	71.09
	Oct. 12	1344	21.78	16.40	1.01	75.30
	Oct. 12	1345	21.48	15.38	....	71.60
	Oct. 12	1346	21.00	15.27	....	72.71
187	Oct. 13	1578	20.00	13.56	....	67.80
208	Oct. 2	839	19.00	12.46	....	65.58
	Oct. 2	840	19.00	12.99	1.28	68.37
209	Oct. 2	832	18.20	13.08	....	71.87
	Oct. 2	833	20.03	14.43	1.17	71.90
	Oct. 2	821	18.27	13.45	....	73.62
	Oct. 2	823	18.00	13.38	....	74.33
	Oct. 2	827	17.43	12.40	....	71.14
212	Oct. 2	828	18.23	13.08	....	71.75
	Oct. 2	829	19.09	14.19	....	74.33
	Oct. 2	830	19.63	14.84	1.38	75.60
	Oct. 2	831	18.60	13.14	....	73.00
	Oct. 13	1516	20.82	15.71	....	75.46
238	Oct. 9	1028	17.48	13.14	2.07	75.17
	Oct. 9	1030	18.88	12.63	....	66.90
	Oct. 5	981	17.82	13.28	....	74.52
	Oct. 5	982	19.05	13.84	....	72.65
	Oct. 5	983	18.20	12.04	....	71.10
	Oct. 5	984	19.50	14.65	1.04	75.13
	Oct. 5	985	18.70	13.44	....	71.87
	Oct. 5	994	18.70	13.25	....	70.86
	Oct. 5	995	18.20	12.88	....	70.77
239	Oct. 8	1002	19.00	13.47	....	70.89
	Oct. 8	1005	19.10	14.00	....	73.30
	Oct. 8	1008	18.30	12.98	....	70.93
	Oct. 8	1010	20.14	15.24	1.55	75.67
	Oct. 8	1011	19.17	13.42	....	70.01
	Oct. 8	1012	19.84	15.17	2.29	76.46
	Oct. 8	1014	18.64	13.55	....	72.69
	Oct. 8	1015	18.70	14.13	....	75.56
	Oct. 8	1020	18.77	13.25	....	70.59
	Oct. 8	1021	19.17	13.28	....	69.27
	Oct. 8	1023	19.20	13.50	....	70.78
	Oct. 16	1815	20.00	13.71	....	68.55

Plots No. 153 and 184 gave some of the best results, the latter especially giving a great many individuals with a high sugar content; there was a great deal of variation in type, however.

ANALYSES OF VARIATIONS IN STANDARD VARIETIES.

The following table gives the results of analyses of individuals canes which were taken from the plots of some standard varities, and which showed some desirable variation from the type of the variety. The variations chosen were in the line of the improvement of the variety. For example, the variations of Honduras were individuals which ripened earlier than the rest of the plot; those of the Link's Hybrid were canes that showed more or less freedom from the faults of the variety, etc. As with the crosses, the analyses given are the chosen ones of a large number of analyses, for none of the canes which showed simply an improvement in external characters were saved unless they showed at the same time a good content of sugar and a high coefficient of purity. The samples in which glucose was determined are the individuals chosen for future planting.

Analyses of variations in standard varieties.

HONDURAS.

No. of plot.	Date.	No. of analysis.	Degree Brix.	Sucrose.	Glucose.	Coefficient of purity.
				Per cent.	*Per cent.*	
225-6	Sept. 3	203	14.43	9.07	1.83	62.28
	Sept. 3	266	13.53	8.46	2.37	62.53
	Sept. 3	267	13.47	8.16	....	60.58
	Sept. 17	306	17.84	12.14	.79	68.05
	Sept. 17	313	18.72	12.59	1.18	67.75
	Sept. 17	316	18.33	12.28	2.74	66.99
	Sept. 20	330	15.35	11.19	....	72.00
	Sept. 20	338	18.28	12.70	2.65	69.47
	Sept. 20	339	14.78	9.94	....	67.25
	Oct. 20	2110	19.10	13.37	....	70.00
	Oct. 20	2117	20.00	14.05	....	70.25
	Oct. 20	2118	20.00	14.90	1.02	74 50

WAUBANSEE.

230	Sept. 21	340	18.32	13.04	.99	71.18
	Sept. 21	341	17.52	12.07	.81	68.89
	Sept. 21	342	15.31	9.57	....	62.47
	Sept. 21	344	14.85	9.74	....	65.59
	Sept. 21	347	15.85	10.19	....	64.29
	Sept. 21	393	15.35	9.58	1.13	62.41

WHITING'S EARLY.

*Analyses of variations in standard varieties—*Continued.

LATE-ORANGE.

No. of plot.	Date.	No. of analysis.	Degree Brix.	Sucrose.	Glucose.	Coefficient of purity.
				Per cent.	*Per cent.*	
89	Sept. 25	492	16.68	10.73	....	64.33
	Sept. 25	498	18.18	12.07	....	66.39
	Sept. 25	500	18.78	13.64	2.94	72.63

EARLY ORANGE.

68	Sept. 28	520	17.91	12.89	....	71.97
	Sept. 28	521	17.21	14.42	.76	83.79
	Sept. 28	522	16.31	14.43	....	88.47
	Sept. 28	529	15.83	10.21	....	64.50

WHITE MAMMOTH.

67	Oct. 3	867	19.45	13.29	....	68.33
	Oct. 3	868	19.00	13.81	....	72.68
	Oct. 3	871	18.62	14.12	....	75.83
	Oct. 3	893	19.20	14.04	1.37	73.13
	Oct. 3	882	19.00	13.86	....	72.95

CHINESE.

232	Oct. 3	901	18.58	13.36	....	71.91
	Oct. 3	907	18.70	14.44	....	77.22
	Oct. 3	909	18.70	13.39	....	71.60
	Oct. 3	910	19.85	14.35	....	72.29
	Oct. 3	911	19.08	14.79	1.06	77.52
	Oct. 3	912	17.35	12.55	....	72.33
	Oct. 3	954	18.24	12.95	....	71.00
	Oct. 3	955	19.84	14.50	....	73.08
	Oct. 3	962	19.19	13.76	....	71.70
	Oct. 3	973	19.17	14.10	....	73.55

LINK'S HYBRID.

0	Oct. 3	924	20.70	15.50	....	74.88
	Oct. 3	932	19.20	14.15	....	73.70
	Oct. 3	937	21.38	15.60	....	72.97
	Oct. 3	938	20.18	15.24	....	75.52
	Oct. 3	941	18.51	13.57	....	73.31
	Oct. 3	942	20.20	14.43	....	71.44
	Oct. 3	944	21.20	15.81	....	74.58
	Oct. 3	949	22.31	16.93	.55	75.89

The unnamed plots also contained a great many interesting variations, selections from which are given in the following table:

Analyses of variations in the unnamed plots.

No of plot.	Date.	No of analysis.	Degree Brix.	Sucrose.	Glucose.	Co-efficient of purity.
				Per cent.	*Per cent.*	
0	Sept. 24	421	15.12	9.18	......	60.71
	Sept. 24	428	18.67	13.84	.80	74.13
12	Sept. 22	416	16.58	10.50	......	63.87
	Sept. 22	417	15.28	10.00	......	65.45
	Sept. 22	418	16.08	11.00	......	68.41
33	Sept. 25	505	18.70	12.48	......	60.74
	Sept. 25	506	18.48	12.02	......	65.04
	Sept. 25	508	16.28	9.48	......	58.23
	Sopt. 25	509	19.48	13.17	2.61	67.61
	Sept. 25	511	19.22	12.56	......	65.35
37	Sept. 22	396	19.60	14.20	1.04	72.76
	Sept. 22	398	18.20	12.94	......	71.10
	Sept. 22	401	18.10	12.00	......	70.11
44	Sept. 21	380	17.75	11.00	......	62.48
	Sept. 21	383	18.42	12.52	......	67.97
	Sept. 21	385	17.15	10.00	......	58.31
	Sept. 21	387	18.39	13.01	......	70.74
	Sept. 21	389	16.22	13.13	......	80.95
45	Sept. 22	402	17.70	12.38	......	69.94
	Sept. 22	406	19.60	13.92	.44	71.02
	Sept. 22	407	17.24	12.07	......	70.01
	Sept. 22	408	17.64	12.29	......	69.07
	Sept. 22	409	17.74	12.14	......	68.43
	Sept. 22	410	20.74	14.86	.37	71.05
	Sept. 22	411	16.74	11.85	......	70.79
	Sept. 22	413	16.28	11.48	......	70.52
46	Sept. 21	357	16.90	13.00	......	76.92
	Sept. 21	361	20.60	14.83	.70	71.99
	Sept. 21	364	19.62	13.47	1.54	68.65
	Sept. 21	365	20.19	14.79	.76	73.25
	Sept. 21	366	20.75	15.14	.77	72.90
	Sept. 21	371	18.28	12.63	......	69.09
	Sept. 21	372	19.15	13.74	.59	71.75
	Sept. 21	373	19.02	14.67	.64	73.64
	Sept. 21	374	19.35	13.08	1.07	70.70
	Sept. 21	377	20.78	14.27	2.09	68.67
	Sept. 25	514	18.31	12.36	2.30	67.50
48	Sept. 4	290	18.44	13.25	1.01	71.85
	Sept. 4	292	17.44	11.09	......	68.75
	Sept. 4	294	17.74	12.71	......	71.63
	Sept. 4	295	17.52	12.04	......	68.72
50	Sept. 21	351	17.60	11.53	......	65.51
	Sept. 21	356	16.40	12.22	......	74.51
	Sept. 24	430	17.72	12.20	......	68.85
	Sept. 24	432	16.68	11.71	......	70.20
	Sept. 24	438	17.68	12.78	......	72.28
	Sept. 24	439	16.48	11.77	......	71.42
	Sept. 24	440	18.28	13.24	......	72.43
	Sept. 24	444	18.48	13.57	.95	73.43
	Sept. 24	446	17.48	11.86	......	67.85
	Sept. 24	447	17.01	12.04	......	70.78
	Sept. 24	449	18.81	13.50	......	71.77

IV. EXPERIMENTS IN THE SELECTION OF SEED FROM INDIVIDUAL CANES SHOWING A HIGH CONTENT OF SUGAR.

VARIABILITY OF SORGHUM CANES IN THEIR CONTENT OF SUGAR.

As might be expected of a plant which varies so much in the outward character of its individuals, sorghum canes vary greatly in the chemical composition of their contained juices. Even in canes of the same varieties, showing uniform outward characters and of uniform appearance and development, great differences will be found in the composition of the juice from individual canes. In fact the variation in this respect seems much greater and more persistent than in the outward appearances of the plant. When the variety itself is not uniform, and the variations due to mixed races are added to the variations of individuals, the most remarkable extremes are produced. This can be seen by examining the analyses of individual canes of crosses given in the section on experiments with crosses, from which the following table is selected, to illustrate the possible differences between different canes growing in the same plot. The canes were selected from a plot of Honduras, which showed fairly uniform character, in the endeavor to obtain early ripened seed of that variety, and probably some were not so well matured as others, though the seed from all was perfectly hard.

Polarization of selected canes from Honduras.

No.	Degree Brix.	Sucrose.	No.	Degree Brix.	Sucrose.
		Per cent.			Per cent.
1.........	6.93	.20	11.........	14.15	8.25
2.........	14.43	9.07	12.........	17.05	11.41
3.........	13.53	8.46	13.........	15.88	10.92
4.........	13.47	8.16	14.........	15.34	9.33
5.........	10.47	4.91	15.........	15.34	7.51
6.........	14.40	7.40	16.........	15.54	6.50
7.........	11.85	5.78	17.........	16.67	11.53
8.........	10.04	1.51			
9.........	11.65	5.24	Highest	....	11.53
10.........	....	.10	Lowest .	....	.10

The following table shows the variation of individuals in a well-established and uniform variety. They were selected with this end in view from a remarkably uniform plot of Early Amber, and a particular effort was made to have the canes as nearly of the same size and general appearance, the same maturity, and the same conditions of growth— all taken from the same row.

Polorization of average canes from Early Amber.

No.	Degree Brix.	Sucrose.	No.	Degree Brix.	Sucrose.
		Per cent.			Per cent.
1............	15.50	10.80	10.........	17.44	11.99
2............	15.70	12.02	11.........	14.94	8.08
3............	14.50	7.54	12.........	17.74	12.71
4............	18.00	12.78	13.........	17.52	12.04
5............	16.74	10.36	14.........	17.32	10.53
6............	14.74	8.58	15.........	17.32	10.88
7............	15.44	9.58			
8............	18.44	13.25	Highest	....	13.25
9............	17.24	11.61	Lowest .	....	7.54

While the difference is not so great as in the previous table, it will be seen that there is a difference of nearly 6 per cent. of sucrose between the richest and poorest canes in fifteen samples.

Even in the highly improved and well established varieties of sugarbeets this variation in the composition of individuals occurs, as will be seen in the following table taken from Stammer,[*] which shows analysis of individual beets taken from the same row.

Polarization of German sugar-beets.

No.	Weight of beet in grams.	Degree Brix.	Sucrose.	Apparent purity.
			Per cent.	
1	350	18.1	14.0	8?.3
2	760	15.7	13.0	82.4
3	640	16.0	12.8	79.7
4	635	15.3	12.8	83.7
5	585	15.3	12.4	81.1
6	650	16.4	13.0	78.8
7	690	15.8	13.8	87.2
8	290	16.5	13.1	70.5
9	532	19.0	17.1	90.0
10	660	16.2	13.5	83.0
Highest			17.1	
Lowest			12.4	

From this it appears that these individual beets showed nearly as great variations as the Amber canes, though from the differences in the weights of the beets it is evident that they were selected at random, with no special effort to obtain average samples, as was the case with the canes.

DIFFICULTIES IN THE SELECTION OF SEED ACCORDING TO CONTENT OF SUGAR IN THE CANE.

It is much more difficult to select the best individuals of a sugar-producing plant than of plants raised for other purposes, in which the relative merit of the individuals can be seen by outward appearances. There are no known reliable outward signs which indicate that a certain cane contains more sugar than the others. In a garden one can select the finest vegetables, in the orchard the finest fruits, in the grain fields the finest ears of corn or of wheat, by the eye or by weight, or by very simple tests. But sugar is inside the canes, mingled with other substances. The weight of the canes or their appearance is not a reliable measure of the sugar which they contain. Handsome canes may contain but little sugar; canes inferior in appearance may yield sugar well. The sense of taste is not a reliable test, for the sugar in the juice is masked by other substances. A sugar-cane which shows by analysis 12 per cent. of sugar tastes much sweeter than a sorghum-cane which shows 15 per cent.

The sorghum plant will be improved but slowly if selections of seed are made only by the size or weight or appearance of the canes, or by simple selections of the finest appearing seeds.

In 2,000 analyses and polarizations of cane juice made at this station there were no reliable and constant outward marks observed by which the canes which contained most sugar could be selected. The degree

* Lehrbuch der Zucker-fabrikation, von Dr. K. Stammer Braunschweig, 1887, p. 150.

of maturity was the only sign, and selections of the richest canes can not be made by that.

When the sugar-beet growers attempted to improve the sugar beet they met with the same difficulty. They were well aware that the hereditary principles which are known to apply to animals also apply to plants. They knew that the individual beets which actually contained more sugar than the others should be saved for planting. But the characteristic points of beets which are rich in sugar vary, so that they are not reliable guides in selecting beets for seed. Knauer invented a machine which separated beets in piles according to their weight, in order to select the heaviest, not the largest, beets for seed. And beets were placed in a solution of salt-water of a certain density; the beets which sank were saved for seed. These methods were only adapted to rough selections. To Vilmorin is due the credit of introducing the methods by which the sugar-beet has been so wonderfully improved. He observed that a cylindrical piece could be taken from each beet without injury to the plant. These sample pieces were separately tested to determine their value in sugar manufacture, and only the beets which were proved to contain more sugar than the others were saved for seed. To show the zeal with which the work of improving the sugar-beet was done, it is only necessary to say that at the Paris Exposition of 1878 there were twenty exhibitors who claimed to have produced improved varieties of the beet. Deprez et Fils of France had an agricultural laboratory with facilities for making 2,000 analyses of beets daily. With the assistance of Professor Viollette they produced three important new varieties of the sugar-beet, which are known as "Improved Deprez," 1, 2, and 3.

It is evident that the sorghum industry should profit by this experience of the beet industry, and that sorghum seed should be saved only from individual canes which yield well in sugar.

ADVANTAGES POSSESSED BY SORGHUM OVER OTHER SUGAR-PRODUCING PLANTS IN THE SELECTION OF SEED.

Sorghum has advantages over both the sugar-cane and the sugar-beet in selecting seed from the best individuals, and it can reasonably be expected that its improvement could be made much more rapidly than has been the case with the former. In the first place the sugar-beet is a biennial plant, requiring two years to produce its seed; sorghum is an annual, requiring but one year to mature its seed, so that its progress should be twice as rapid; then the sorghum is unique among sugar-producing plants in that its seed may be separated entirely from the cane and the latter analyzed, giving exactly the worth of the individual which produced the seed, without injury to the seed itself. This is a vast improvement over the tedious method that must be pursued with the beet, of cutting out a portion of the root for the purpose of analysis. Such a cylinder can not represent the quality of the whole root with entire accuracy, and there is ground for supposing that it somewhat impairs it for the production of seed the next year, although the originators and those practicing the method claim it does not. Certainly the analysis of the entire portion of the plant which is used for sugar-making purposes, as is possible in sorghum, is greatly superior. The sugar-cane is at a tremendous disadvantage in this respect, and this is undoubtedly one reason why it has fallen behind the beet in the struggle for supremacy as a sugar-producing plant. Being propagated by eyes, or suckers, there is no way of obtaining an analysis of the cane without injuring it for seed purposes.* The result has been that the plant has deteriorated rather than improved, while the sugar-beet has steadily advanced in quality.

Surely it would be criminal folly on our part if we failed to avail ourselves in the sorghum industry of the advantages naturally possessed by the plant, and of the lessons taught us by the experience of others with the beet and the cane.

METHOD OF WORK EMPLOYED AT THE STERLING EXPERIMENT STATION.

Owing to the pressure of work at this station the past campaign, and the attention given the crosses, the selection of seed from the best individual canes of the established varieties was not instituted until late in the season, and could not be carried out on the earlier varieties. The selection should properly be made, of course, at the maximum of maturity of the cane. The plan of work was as follows: A large number of canes were selected from the plot, care being taken that those selected should show no outward faults of form, and should be average canes in size, of good healthy appearance. A large number of such canes were brought in to the station barn and laid out in serial order, the heads cut off, a label with number attached to each, and a corresponding number placed on a receptacle to contain the juice. Two men were kept busy turning the hand-mill, while a third kept the juices in proper order. As soon as the juices were obtained they were poured into hydrometer jars, and when they had stood long enough to permit of the escape of the air bubbles, their density was taken roughly with a spindle. If the reading did not come up to a certain standard the juices and corresponding seed-heads were rejected. The standard used depended upon the richness of the variety of cane from which the selections were made, being placed at $20°$ or even $21°$ Brix for very rich

---

* Professor Stubbs has proposed to split the cane, using one-half for analysis, and the other for planting. Of course there would be considerable difficulty in preserving the split cane, and there is no record of its ever having been attempted. It would seem more feasible to cut a short section, containing one eye, from a stalk for planting, and make the analysis on the remaining portion of the stalk.

The success of Professor Harrison in the Barbadoes in producing sugar-canes from seed (Royal Gardens, Kew, Bulletin of Miscellaneous Information, 1888, No. 24, p. 291), would seem to give hopes for the improvement of the plant in the way of new varieties, and the present method of propagating the plant from any kind of individuals that may be most convenient should receive equal attention; it is simply barbarous.

varieties like the Links Hybrid. The few juices which passed the test were sent to the laboratory for complete analysis, and the corresponding seed-heads carefully preserved. From the complete analyses, still further selections were made, so that ultimately a few seed-heads were saved of canes showing great richness and purity of juice. From 500 to 1,000 canes could be tested in this way in a day. Some of the canes obtained by this method of selection were very rich in sugar. The following instances serve to show this.

A plot of Links Hybrid, of which the highest analysis from average samples had been 14.09 per cent. sucrose, gave on selection from about 500 canes four which went over 15 per cent.

Another plot of the same variety, showing by analysis of an average sample 12.24 per cent. sucrose, gave by selection from 500 canes three which had over 16 per cent. sucrose in the juice.

An average sample of a plot of Liberian cane gave 14 per cent; 500 canes were taken from different parts of the plot and one cane gave 17.59 per cent. sucrose in the juice; three gave over 16.5 per cent., and twelve over 15.5 per cent.

An average sample of the Planters' Friend, a new variety from Australia, gave 11.63 per cent. sucrose; selections from 1,000 canes gave three which contained over 15 per cent. sucrose in the juice. Such instances might be multiplied, but sufficient evidence has been given to show the possibilities in this method of improvement. The selections have all been preserved, and can be planted and observed another year, if means are afforded the Department for carrying out the work.

Analyses of selected single canes from standard varieties.

LIBERIAN.

No. of plot.	Date.	No. of analysis.	Degree Brix.	Sucrose.	Glucose.	Co-efficient of purity.
				Per cent.	*Per cent.*	
222	Oct. 3	842	18.46	12.96	....	70.21
	Oct. 3	843	18.36	13.29	....	72.39
	Oct. 3	845	18.90	13.47	....	71.27
	Oct. 3	847	19.32	13.30	....	68.84
	Oct. 3	856	19.30	14.22	1.31	73.68
	Oct. 17	1944	21.13	15.71	....	74.35
	Oct. 17	1945	20.73	15.64	....	75.45
	Oct. 17	1948	20.83	15.49	....	74.36
	Oct. 17	1950	22.91	16.42	2.28	71.67
	Oct. 17	1951	22.41	16.20	1.22	72.29
	Oct. 17	1953	21.71	17.69	1.97	81.48
	Oct. 17	1955	22.28	15.74	....	70.65
	Oct. 17	1956	20.80	15.30	....	73.56
	Oct. 17	1960	21.32	15.12	....	70.92
	Oct. 17	1961	21.62	15.34	....	70.95
	Oct. 17	1962	21.71	15.06	....	69.37
	Oct. 17	1967	21.91	15.00	....	68.46
	Oct. 17	1968	22.41	16.49	....	73.58
	Oct. 17	1969	21.31	15.70	....	73.67
	Oct. 17	1983	21.58	16.75	....	77.62
	Oct. 17	1985	21.58	15.74	....	72.94
	Oct. 17	1993	22.38	15.93	....	71.18
	Oct. 17	1996	21.92	14.62	....	66.70
	Oct. 17	2000	21.62	14.75	....	68.22
	Oct. 17	2002	22.88	15.66	....	68.44
73	Oct. 16	1805	18.70	13.34	1.84	71.34
72	Oct. 17	1900	19.72	14.18	....	71.91
	Oct. 17	1905	19.88	14.37	....	72.28
	Oct. 17	1907	19.68	13.99	....	71.09

*Analyses of selected single canes from standard varieties—*Continud.

LINKS HYBRID.

No. of plot.	Date.	No. of analysis.	Degree Brix.	Sucroso.	Glucoso.	Co-effi-cient of purity.
				Per cent.	*Per cent.*	
	Oct. 18	2007	20.77	15.30	....	73.66
	Oct. 18	2009	21.58	15.34	....	71.08
	Oct. 18	2010	21.08	15.57	....	73.86
	Oct. 18	2015	21.10	15.18	....	71.94
	Oct. 18	2020	21.28	15.95	....	74.95
	Oct. 18	2021	21.58	16.11	....	74.65
	Oct. 18	2022	21.17	15.85	....	74.87
	Oct. 18	2029	21.12	15.92	.60	75.38
	Oct. 18	2030	21.70	15.84	.92	73.00
	Oct. 18	2038	21.50	15.52	1.09	72.19
	Oct. 18	2048	21.53	16.51	....	76.61

EARLY ORANGE.

[Selections taken from large cane-field.]

	Oct. 18	2040	22.18	17.05	.67	70.87
	Oct. 18	2041	21.46	15.26	....	71.11
	Oct. 20	2102	21.68	14.88	....	68.63

EARLY ORANGE.

74	Oct. 17	1940	20.70	15.01	....	72.51
	Oct. 17	1941	20.50	13.58	....	66.24

PLANTERS' FRIEND.

214	Oct. 16	1820	21.70	14.71	....	67.79
	Oct. 16	1831	22.00	15.55	.82	70.68
	Oct. 16	1839	19.70	14.16	....	71.88
	Oct. 16	1842	20.10	14.94	....	74.33
	Oct. 16	1844	20.00	14.00	....	70.00
	Oct. 16	1853	21.58	14.53	....	67.33
	Oct. 16	1856	20.20	14.18	....	70.20
	Oct. 16	1857	20.60	15.41	1.60	74.81
	Oct. 16	1863	20.28	14.51	....	71.55
	Oct. 16	1877	20.60	14.68	....	71.26
	Oct. 16	1880	21.00	14.86	....	70.76

CHINESE.

215	Oct. 13	1508	19.68	14.71	1.23	74.75

The following table gives the highest analysis obtained in each of five varieties by selection:

Highest analysis of single canes by selection from standard varieties.

Variety.	Date.	No. of analysis.	Degree Brix.	Sucroso.	Glucoso.	Co-effi-cient of purity.
				Per cent.	*Per cent.*	
Liberian..........	Oct. 17	1953	21.71	17.69	1.97	81.48
Early Orange......	Oct. 18	2040	22.18	17.05	.67	70.87
Links Hybrid	Oct. 18	2030	21.70	15.92	.69	75.38
Planters' Friend...	Oct. 16	1831	22.00	15.55	.82	70.68
Chinese..........	Oct. 13	1508	19.68	14.71	1.23	74.75

V. Experiments in Improvement by Methods of Cultivation.

It is a rule in agricultural science that to obtain the best results the individual plants must be given the most favorable conditions possible for full development.

In the effort to improve the sorghum plant methods of cultivation will play an important part. Very little attention has been paid heretofore to this subject, the cheapest and easiest methods being followed; and the sorghum crop has had about the same cultivation as is given to the corn crop. In the work at this station no very extensive experiments could be made on different methods of cultivation, but a number of practical points were evolved, which may be stated as our views on the best methods to be followed, without going into details as to the evidence upon which the conclusions were based.

It is desirable in growing cane for sugar manufacture, that as nearly as possible all of the plants in one field should ripen at one time. If in one row there are some canes fully ripe and other canes immature, it will not be easy to harvest the canes at the time when each contains its maximum of sugar. It is a point of advantage to have all come up at the same time. This can best be accomplished by planting the cane on freshly plowed land the same day the land is plowed, and by being careful to cover the cane seed at a uniform depth with earth. This insures as uniform a start as possible for the canes, and while it may seem a trifling matter it often materially affects the results.

After the young plants have come up a serious problem arises, and that is, how to cultivate the plants, to pulverize and loosen the soil, and to destroy the weeds without injuring the roots on which the development of the plants depends.

Great injury is done to the roots of canes when the cultivator works deep and close to the plants after they have attained considerable size. This injury is perhaps greater than most persons suppose. It appears to be proved by a very simple experiment. If the roots of a hill of cane are cut all around the hill with a spade at a distance of 6 inches from the canes to a depth of 6 inches from the surface, when the plants are 4 inches high, and if this process is repeated once a week until the canes are 4 feet high, the canes thus treated will be found to ripen later and to be inferior in all respects. In wet seasons the injury is not so great as in dry, but injuries are caused to growing plants by the cultivator as with the spade.

To avoid destroying and mutilating the roots of the growing canes, it seems better to give deep and close cultivation while the plants and their roots are small, and when the first cultivation is given to use long and narrow shovels, which work near the canes, and with a slow and steady team give close and deep and thorough cultivation before the rootlets are expanded sufficiently to be injured by such cultivation.

In the succeeding cultivations " shallow shovels, " that is, shovels having such form that they do their work at or near the surface of the soil, should work near the plants, while deeper cultivation may be had at a distance from the plant which the roots have not reached.

The form of shovel preferred in the experiments at this station is known as the "Eagle's Claw." It consists of eight small shovels, which are attached to the beams of a two-horse cultivator, four shovels working on each side of the row of cane. The form of these shovels is such that they do not enter the soil deeply; they thoroughly pulverize all the surface soil and destroy weeds, and work close to the growing plants with little injury to the roots.

We have alluded to these points because we believe the yield of sugar is often materially lessened by injuring the roots of the canes. Mutilation of the cane plants above the surface of the soil is known to produce a lessened yield of sugar, and injuries to the cane plants below the surface doubtless decrease it also. Many cane-growers, as they "lay by" their cane crop, or finish the cultivation and see its deeply and closely cultivated canes free from weeds, do not realize that while destroying weeds they nearly destroyed their cane plants, and while working for their canes they were working against them and against the yield of sugar.

DISTANCE APART AT WHICH CANES SHOULD STAND IN THE ROW.

This is a subject which has attracted considerable interest among sorghum-growers lately. Mr. Hughes obtained last year the highest yield of sugar per acre ever reported for sorghum. According to his statement, "this was occasioned by carefully planting the hills closer and giving it good attention, together with favorable rains."* As a contribution to the solution of this important question, the following analyses may be recorded.

Two experiments were made with different plots of cane, both of which had been planted with drill. The planting had been rather uneven, and some rows were much thicker than others. A thick and a thin row in each plot were chosen, the canes counted and cut for a considerable distance, which was measured, and the whole run through the mill. The number of canes divided into the length of row cut gave the average distance apart of the canes in the row, and from the weight of the whole sample the yield per acre was calculated. The following table gives the results of the analyses:

	Yield, tons per acre.	Degree Brix.	Sucrose.	Glucose.	Co-effi-cient purity.
Experiment No. 1, Early Amber cane:			Per cent.	Per cent.	
Growing 4 inches apart in the row ...		18.20	12.50	2.91	68.5
Growing 7½ inches apart in the row ..		lost	14.01	2.60	
Experiment No. 21, Early Orange cane:					
Growing 3.4 inches apart in the row .	0	17.82	11.73	2.82	65.8
Growing 8.8 inches apart in the row .	7.1	18.70	13.16	2.62	70.4

* Bull. No. 17, Chem. Div., U. S. Dept. Agriculture, p. 68.

These analyses show very decided differences between the two samples. The conditions were in all respects similar, except as to the distance apart of the canes in the row, and the large samples taken diminished the possible error of sampling, so that considerable reliance may be placed upon the results. It will be seen that in both experiments the canes which were thin in the row were much better in quality than those which stood closely together; the content of sucrose is higher, of glucose lower, and the purity is greater. It is evident that close planting, while it increases the tonnage, diminishes the yield of sugar per ton. Of course there is a proper mean between too close planting on the one hand and too thin planting on the other, and this subject is worthy of more attention and discussion than has previously been given it. It is probable that the distances at which canes should be planted vary to some extent with the varieties. For instance, it would seem that the small canes of the Early Amber do not require so much space as the much larger canes of the Honduras, and it also seems that soils and climate may require the distances between the canes to vary. For instance, it is well known that corn is planted much closer in the North than in the South.

VI. Miscellaneous Experiments and Results of Observations.

EFFECTS OF FROST.

The effect of a light frost upon sorghum cane has always been a mooted question, some holding that it is not injured by a frost which only kills the leaves, but rather has the effect of ripening the cane. It seems reasonable to suppose that it does affect it unfavorably, however, as it kills the leaf and stops further growth and vitality in the plant. The question is an important one, for it is quite common to have a slight frost quite early in the season. A few observations were made on this point at this station the past season. The first frost occurred on the night of September 27. On October 5, about a week afterwards, when the effects of the frost were plainly perceptible, the different plots were examined to see if any observations of importance could be made. The more immature varieties seemed to have resisted the action of the frost better than those which were more matured; the Honduras, for instance, holding the bright green of its leaves, almost without exception. Some varieties appeared to have resisted the action of the cold much better than others, giving some ground for the hypothesis that this might prove a constant characteristic. Other plots, again, showed some spots that were almost entirely untouched by the frost, while in other spots the leaves were quite dead, the differences being doubtless due to different conditions of evaporation from the soil. These plots seemed to offer an opportunity for comparative analyses of frosted and unfrosted canes. Large samples were taken of both kinds, taking all the care possible to have them comparable in all respects, except as to the frosted and unfrosted condition. The results are given in the following table:

Analyses of frosted and unfrosted canes of the same plot.

Variety.	Not frosted.				Frosted.			
	No. of analysis.	Degree Brix.	Sucrose.	Glucose.	No. of analysis.	Degree Brix.	Sucrose.	Glucose.
			Per cent.	*Per cent.*			*Per cent.*	*Per cent.*
Waubansee........	468	16.32	11.71	.91	469	16.72	10.75	1.50
Waubansee (another plot)	498	15.87	11.10	.92	499	15.07	9.96	1.50
Red Liberian	471	18.80	13.52	1.67	479	18.20	12.75	1.38
Red Liberian (another plot)	481	19.20	13.56	1.42	482	18.15	10.26	2.93
Plot No. 67........	474	16.33	11.90	1.26	475	15.01	10.30	1.39
Enyama	477	18.10	12.74	1.47	478	15.70	10.96	1.71
Average........	....	17.44	12.42	1.27	....	16.14	10.83	1.73
Coefficient of purity	....	....	....	71.2	....	....	....	67.1

It will be seen that in every case the juice from the frosted canes was quite inferior. The average of the six different plots shows the juice from the frosted cane was lower in solids, lower in sucrose, higher in glucose, and of less purity than the juice from the canes which had been but little touched by the frost, as shown by the leaves being fresh.

While not sufficient in number to establish the point, these analyses seem to show that sorghum cane deteriorates after the leaves are killed by frost.

ANALYSES OF SAMPLES FROM ARKANSAS.

The capabilities of Arkansas as a sorghum-growing State have never been very extensively investigated. The Sterling Sirup Works received this fall a bundle of cane from one of the "prairie counties" of Arkansas, and the different samples were analyzed at the station, with the following results:

Analyses of canes from Arkansas.

Variety.	No. of analysis.	Degree Brix.	Sucrose.	Glucose.	Coefficient purity.
			Per cent.	*Per cent.*	
Texas Red	545	20.25	13.80	2.84	68.1
Honduras	546	20.25	3.68	8.47	18
Chinese..........	547	18.25	11.05	5.24	61
Orange	548	19.25	14.24	2.23	74

As a general rule samples of sorghum sent from one point to another by express are so much inverted when they arrive at their destination that the analyses are worthless; and then when samples of a few canes are selected by persons not familiar with the plant, the largest and finest-looking canes are chosen, which generally give a lower per cent. of sugar than average-sized canes. In view of these facts, the above analyses make a remarkably fine showing for the locality which produced the canes. The samples all consisted of quite large fine canes, but still gave a good analysis. The sample of Texas Red was a tremendously large cane. The samples of Honduras and Chinese had evidently inverted slightly, the others very little.

Another lot of samples received by the Sirup Works from Thomas Leslie, Stuttgart, Ark., consisted of the following varieties : Goose-neck, Honduras, and Orange. As the analysis showed all to be badly inverted, it is not worth while to give the results.

ANALYSES OF SUGAR-BEETS.

A few samples of sugar beets were brought into the station for analysis by farmers living near town. They were grown from imported seed which had been distributed in western Kansas by Mr. Claus Spreckels, in the spring of 1888. The following table gives the results :

Analyses of sugar beets.

From—	No. of analysis.	Degree Brix.	Sucrose.	Glucose.	Ash.	Coefficient of purity.
			Per cent.	*Per cent.*	*Per cent.*	
Mr. Rimmers.......	459	13.18	8.36	.43	....	63.4
Do	464	14.32	9.61	.29	1.89	67.0
Mr. Stubbs.........	463	11.22	8.92	.30	2.19	62.7
Mr. Schlichter......	465	14.41	9.75	.24	2.04	67.6

These analyses seem to furnish evidence to the effect that this part of Kansas is better suited to the growth of sorghum than the sugar beet. None of the samples above show a sufficiently high percentage of sugar to make them available for profitable sugar manufacture, and the high percentage of ash shown is remarkable ; it is doubtless due to the highly saline character of the subsoils in this locality.

ANALYSIS OF FROZEN CANE.

On the night of October 19, most of the cane still standing in the field was frozen. In continuation of the work on development a sample was taken early in the morning from the plot of Link's Hybrid, and when the canes were run through the mill they were found to be partially frozen. The juice was analyzed, however, and the analysis is given here, together with the analysis of the juice from another sample from the same plot taken later in the day, after it had " thawed out."

Analysis of frozen cane.

Description.	No. of analysis.	Degree Brix.	Sucrose.	Glucose.	Coefficient of purity.
			Per cent.	*Per cent.*	
Juice from sample taken while frozen.........	650	27.10	19.43	1.81	71.7
Juice from sample taken after thawing out....	651	18.26	13.45	.90	73.7

This analysis is inserted more as a matter of curiosity than anything else. It shows simply that part of the water in the juice was frozen, so that the juice expressed was more dense than ordinarily. It might also be used to illustrate the imperfection of the present method of determining the composition of a cane by the analysis of the juice expressed from it by a mill. Such analyses are always subject to the variations of the degree of extraction by the mill, the dryness of the cane, etc. Of course the removal of part of the water from the juice on account of the cane being frozen would not often occur, but a loss of water by drying would also have the effect of increasing the density of the juice extracted. It is to be hoped that methods will be perfected that will admit of the proper sampling of the cane itself, and the direct determination of the sugar.

SIZE OF SORGHUM SEED.

The size and weight of sorghum seed varies greatly in different varieties, and in different individuals of the same variety. Professor Henry found 27,680 seeds to the pound of Wisconsin Amber. Dr. Collier found 19,000 in Virginia Amber. In a sample of the Early Amber seed grown at this station there were 20,200 seeds to the pound. In a sample of the New Orange variety there were 21,760 seeds to the pound. In a sample of Doura (non-saccharine) there were 10,480 seeds to the pound. This variety has the largest seed of any grown here. In an average sample of the Red Liberian variety there were 31,400 seeds to the pound. This has the smallest seed of any variety grown here. In a pound of seed of the same variety, selected for large size, there were 21,800 seeds, one-third less than the average sample.

The vigor of the young sorghum plants in the first weeks of their existence corresponds closely to the weights of the seeds which produced them.

It seems evident that more vigorous plants can be procured by selecting seeds which are above the average size. The Liberian, for instance, produces very small seeds, and these produce very small and slow-growing plants while they are young, although they eventually produce large and handsome canes. It will be noticed that the sample of larger seeds selected from the Liberian had the same weight as average seeds of other varieties.

It is to be supposed that these larger seeds would produce more vigorous plants than the average seeds of that variety.

Major Hallett found that by selecting the finest grains of wheat he improved the plants and also improved the variety.

Mr. Wilson separated the largest and the smallest seeds of the Swedish turnip; he found that the plants from the largest seeds took the lead and maintained their superiority to the last.

Director Briem made similar experiments upon sugar beets, as follows: [*]

"It is a well-known fact that seeds of different size and weight of any plant will correspondingly develop plants of different size and weight if conditions of life are

* Wiener Landwirtschaftliche Zeitung, 1887, No. 99, S. 703. Agric. Science II, 141.

otherwise equal. To determine the amount of these variations in the sugar beet the author made the following experiment : Six bunches of seed from one mother beet were selected, cultivated separately, and the developed plants transplanted after thirty-seven days in such a manner that each plant had the same space of soil. The crop of beets gave the following results:

Seed bunches.	Plants produced.					
	1	2	3	4	5	6
	Grams.	*Grams.*	*Grams.*	*Grams.*	*Grams.*	*Grams.*
No. 1 contained six seeds	1,160	820	720	240	240	110
No. 2 contained six seeds	598	550	415	400	255	245
No. 3 contained five seeds....	735	685	420	310	105	....
No. 4 contained five seeds....	635	625	500	115	35	....
No. 5 contained four seeds...	580	525	350	335	....	....
No. 6 contained four seeds...	370	350	310	55	....	....

From these numbers will be seen the great difference in the weight of the beets although produced from the same bunch. This illustrates the great variability of the sugar beet in inheriting properties, and suggests the greatest care in selecting seed for culture."

In the work in the selection of the individual canes which contained the highest percentage of sugar it was noticed that almost without exception the seed-heads of these canes were far below the average in size and weight. This will be seen by an inspection of a photograph which was taken, showing the seed-heads which gave the highest analysis in the work this season. It may also be remarked that the non-saccharine varieties are invariably large seed-bearers, and have magnificent seed-heads.

Perhaps the simultaneous production of a large amount of seed and of a high percentage of sugar are incompatible?

When any selection of sorghum seed is practiced at all it is the universal custom to select the largest and finest seed beads, but perhaps this method of selection is better calculated to improve the yield of seed than the yield of sugar. This is a most interesting and important question, and we commend it to future investigators.

CONTINUATION OF THE WORK IN THE IMPROVEMENT OF THE SORGHUM PLANT.

The necessity for the continuance of this work has already been sufficiently pointed out. Even after highly sacchariferous varieties have been produced careful selection of seed will still be necessary in order to maintain a high standard of excellence. Who is to carry on this essential branch of the industry ? In Europe the beet industry is sufficiently extensive to justify large seed concerns in undertaking such work, and some of the largest factories save their own seed. In some of the beet-growing countries the agricultural experiment stations render efficient aid in this direction. In this country the industry is still so young that it can not be expected that private capital will undertake the task of improving the plant. The new factories have so much to contend with that they can not possibly devote the necessary time and expense to it. The agricultural experiment stations, in whose province it would seem to fall, have been but recently established in the sorghum-growing States, and are not fully equipped for such work, besides having their attention taken up by other agricultural products. Yet several of them have already done something in the line of sorghum improvement, and others have announced their intention of doing so. It would seem to be essentially fitting and proper if the Department of Agriculture were provided with authority and means for its continuance.

Whoever it may be that undertakes the work, it is important that they should have the benefit of whatever the experience of the past season at this station has taught ; we think it advisable, therefore, even at the risk of some repetition, to outline in a general way the principles and methods to be pursued in the future conduct of such work. It must be remembered, of course, that we have only the experience of one short season to draw upon, and while many of our ideas are based upon that, and upon analogies in beet culture, some have only the foundation of our own judgment to rest upon.

In selecting sorghum seed the following may be outlined as the general course of procedure :

1. Seed should be selected from the varieties which have proved to be the best adapted to the locality. Those which are defective in any respect should either be thrown out or their faults removed by such crossing or selection as will have that tendency.

2. The seed of these varieties should be selected from the individuals which show the fewest faults of form, the highest content of sugar, and the least content of other substances.

3. The seed from the best individuals should receive such cultivation and fertilization as may be shown by experiment to give the best results in yield of sugar, in proportion to the area of soil covered.

It may seem impossible to carry on these several lines of selection at once; to select seed from the individual canes which yield most sugar, and at the same time to select seed with reference to the physical characters of the canes. But more than one point is always necessarily considered in all plant selections. The faults of form in the beet have been bred out, merely to obtain a form to admit of ready cleansing.

The faulty forms of the sorghum cane have already been pointed out. Seed should never be saved from " tillers," or secondary canes, or supplementary heads, as they would tend to reproduce canes which would produce a second crop of seed.

Photographs of some of the canes selected for future propagation at this station will show how faulty forms inherent to certain varieties may be eliminated. The canes are from a plot of Link's Hybrid. This variety has nearly always proved to be a good sugar-producing variety, and its greatest fault is one of form. The top joint is apt to be very long,

slender, and tapering; and as the seed head is pretty heavy it sways back-
wards and forwards in the wind, and in storms is very apt to "lodge."
Selections made from a rather limited number of canes in which this
tendency was partially eliminated gave individuals which were great
improvements upon the typical cane of the variety. This shows how a
fault may be gradually eradicated by selection of desirable variations.
Again, from a cross of the Links Hybrid with the Early Orange, individuals were obtained which retained most of the desirable qualities
of the former, with its typical seed head, and engrafted upon the stout,
stocky canes of the latter. This shows the breeding out of an undesir-
able quality by crossing. The photographs which show the canes obtained by these two different methods contrasted with typical canes of
the variety, illustrate very graphically the possible progress that can
be made in two generations in the improvement of a variety in form.
Faults in form are so readily seen, that it is much more easy to eradi-
cate them by selection than faults of composition, which can only be
ascertained by chemical analysis. Only such canes, then, should be
taken for analysis as show not only freedom from general faults, but
also a tendency towards elimination of faults of the variety to which it
belongs.

It may be as well to insert here a caution as to the use of crossing.
It has been shown that sorghum is extremely variable, and this fact is a
sort of guaranty that by continued selection improved varieties can be
created, for variation makes selection possible, and selection makes improvement possible, but care should be exercised in making use of this
tendency. There is a well-founded prejudice against "mixed" varieties
of sorghum. The most worthless men, animals, and plants are those
which belong to heterogeneous and indiscriminately mixed races. Bad
qualities are transmitted as well as good. The most of the crosses
grown at this station were worthless. Darwin says, "A variety may be
variable, but a distinct and improved race will not be formed without
selection." After the desired degree of variation in the variety has
been obtained selection should be based upon uniformity rather than
variability, in order that the qualities may become fixed and stable.

The most careful and rigid precautions should be taken against accidental crossing, none being permitted that is not artificially controlled
by methods well known to horticulturists. It would probably be well
to prevent cross fertilization even in plots of the same variety. Inten-
sive cultivation has yet to be tried on the sorghum plant, and perhaps
where there is already so great a range of variability, there is greater
prospect of improvement by selection and self-fertilization than by
crossing. It would certainly be best in crosses, and probably best in
varieties, to plant single plots from but one seed head.

In selecting the seed from the best individuals by analysis of the
juice, not only the percentage of sugar, but also the purity and the
percentage of glucose must be considered. This problem is rendered

easier of solution by the fact, which was pretty generally noticeable in the work at this station, that purity of juice and a low content of glucose generally accompany a high percentage of sucrose. Moreover, it is generally the case, though this is not so constant, that a high density of juice indicates high content of sucrose, low of glucose, and high purity. The following analysis, taken from some of the individual canes which gave the best polarizations illustrates this point.

Analyses of canes showing high percentage of sugar.

No.	Degree Brix.	Sucrose.	Glucose.	Coeffi- cient of purity.
		Per cent.	*Per cent.*	
422	22. 50	17. 18	. 58	76. 4
627	22. 16	17. 05	. 06	76. 9
451	22. 28	16. 93	. 55	76. 0
563	22. 00	16. 93	. 70	77. 0
564	22. 50	16. 85	. 91	75. 0
565	21. 25	16. 33	. 77	77. 0
738	18. 61	14. 98	. 75	80. 5
430	19. 47	14. 52	. 53	74. 2

Rough selections therefore can be made by the hydrometer spindle, throwing out all which do not come up to a certain standard. The selections made in this way may then be polarized, and further selections made by this test, while the final selection should be based upon a complete analysis. Considerable weight should be attached to the purity as a basis for selection, for this is the weak point of sorghum as a sugar-producing plant. It will be seen from the above analyses that these canes were fully equal to tropical canes so far as a high content of sucrose and a low content of glucose are concerned, but the purity is low in proportion. The selection and comparison of canes for seed should be made when the plot has reached its maximum of purity, as nearly as it is possible to ascertain that point. Then the relation of high sucrose content, high density, and purity, etc., is most likely to be normal and constant.

It will be seen that a course of selection, as outlined above, necessitates the making of a great many analyses. Facilities for making a large number of analyses, the more the better, would constitute an essential part of the equipment of a station for the improvement of the plant. But undoubtedly a great deal could be accomplished in selection of seed by the use of the hydrometer alone, where facilities for complete analysis do not exist, until the time comes when seed improvement can be carried out properly by separate stations or establishments.

We believe that every large cane grower should test his canes in this way, and should make selections of seed by the hydrometer test, unless he can use still better methods. Even this simple method of selecting seed would be vastly better than the usual way of merely selecting seed that is ripe and sound. If constantly practiced it would do much to remove the blame of variableness from the sorghum plant, because it would throw out from the seed selections the seed-heads from canes which have weak juice which contains little sugar. English horticulturists call destroying inferior plants "rogueing," and the sorghum plant now needs constant rogueing. This can best be done by throwing out the seed of canes which have weak juice.

We can as yet lay down no rules in regard to the selection of either varieties or individuals with reference to the size or yield of seed. This can only be done when it has been settled beyond a doubt that high saccharine content and purity of juice can co-exist with a large yield of fine seed. Time and experience only can settle this question, for we have no analogies to guide us. The seed is a most important by-product in sorghum; it stands alone among sacchariferous plants in its ability to furnish at the same time both a product of sugar and a crop of valuable cereal grain. Both the quality and quantity of the seed produced vary greatly in different varieties; some of them, such as the Honey-dew, White African, White India, etc., furnish a beautiful white seed; the seed-producing qualities could doubtless be easily improved by selection, and the opportunity thus offered is very tempting; but for the present it seems more rational not to expect nature to honor duplicate drafts upon her treasury; to produce a big crop of seed and a large yield of sugar from the same piece of ground that ordinarily does only the former. While it will never do to attempt to place a limit on the possibilities in the case, much must be done before we can expect to produce a sorghum cane with the sugar content of Links Hybrid or Liberian, combined with the seed-head of Doura or Kaffir corn.

Nothing has been said as yet of a very important element which must be considered in all improvements of a race of animals or plants; that is, the power of the selected individual to transmit its qualities to its descendants. An individual may be ever so rich in good qualities itself, but if it does not possess also the power of impressing its own character upon its posterity it is not the best one to choose for breeding purposes. This point is well set forth in the letter which we append, from the celebrated seed firm of Paris, who have done so much for the improvement of the sugar-beet, and whose historical connection with the introduction of the sorghum plant into this country will lend especial interest to what they have to say with reference to its improvement.

LETTER FROM VILMORIN, ANDRIEUX & CO., ON IMPROVEMENT OF SORGHUM.

PARIS, *November* 6, 1888.

DEAR SIR: Replying to your inquiry as to the best method of improving the sorghum plant, we should think that nobody being more acutely aware than you probably are of what qualities are still lacking in this plant, you must, of necessity, be better than any one else in a position to make the first step towards success; which is, to have a clear and precise perception of the aim to be arrived at, *i. e.*, of the most important features to be added to those already existing in the plant.

Besides, if we draw a correct conclusion from what we gather from your letter, not only has this step been already made, but you expect even to have now ready on hand the necessary materials for going a step further, viz, proceed to the selection of those plants as possess to the desirable degree the very qualities looked for.

Here we must remark, as regards the said selection, that, as far as our experience goes, it does not seem to be always the safest way to systematically discard all the merely satisfactory plants and to give the preference only to those showing some qualities to the highest degree. On the contrary, it has often been the case that specimens of only average value, but otherwise well fitted plants, have proved to afford the surest means of rapidly obtaining a final result.

This applies especially to the most important quality to be secured in the selected plants, which is the capability of fully transmitting their good qualities to their descendants, and as this quality can not well be ascertained at the outset, it is necessary not to be over severe in the first selection, and subsequently to retain only those plants as show this quality to a satisfactory degree and then to make a careful selection amongst these.

During the whole time of these experiments it will be necessary to take the required measures to prevent intercrossing, so that the successive progeny of each individual plant be kept severely by itself, and every hybridization be made impossible, as otherwise, even one accident might be conducive to impart to the plants a tendency to variation, which may make it the more difficult to obtain that lasting constancy or fixity so necessary in the plants that are intended to create a new and large generation liable to improve rather than degenerate. As a consequence, it will also be necesary to provide, from the outset, for a most careful and correct record of the signs, the degree, and progress of each of the qualities recognized in each individual plant selected for future propagation. For it is very important that when selecting stock plants amongst the new generation an accurate and easy comparison of each of these plants may be made with every one of his ancestors, so that the increase gained in constancy or permanency of each character wanted may be surely ascertained, and a headway movement secured with certainty.

Of course the number of series to be studied separately may vary according to circumstances, each being conducted on a somewhat different basis as regards the most prominent qualities noticed in the plants used.

By such means and by never altering, without good reasons, the program once laid down at the beginning, you may expect to bring the desired result more or less rapidly into the domain of established facts. Of course, much depends on the skill shown in the successive selections to be made, on the nature of the plants treated, as also on circumstances.

In reply to your query about publishing this letter, we have only to say that if you are of opinion that others may derive some benefit from reading it, we shall not have the least objection to your publishing it.

We remain, dear sir, yours faithfully,

VILMORIN, ANDRIEUX & CO.

Mr. W. P. CLEMENT,
Sterling Sirup Works, Sterling, Kans.

INDEX.

A.

M.

T.

V.

W.

O

www.ingramcontent.com/pod-product-compliance
Lightning Source LLC
Chambersburg PA
CBHW021810190326
41518CB00007B/529